COMPLETE BOOK OF
FRAMING

An Illustrated Guide for Residential Construction

Scot Simpson

WILEY

EXPANDED

SECOND EDITION

COMPLETE BOOK OF

FRAMING

An Illustrated Guide for Residential Construction

Scot Simpson

RSMeans
FROM THE GORDIAN GROUP®

WILEY

Cover image: © Scot Simpson
Cover design: Wiley

Copyright © 2019 by John Wiley & Sons, Inc. All rights reserved.

Published by John Wiley & Sons, Inc., Hoboken, New Jersey.
Published simultaneously in Canada.

No part of this publication may be reproduced, stored in a retrieval system, or transmitted in any form or by any means, electronic, mechanical, photocopying, recording, scanning, or otherwise, except as permitted under Section 107 or 108 of the 1976 United States Copyright Act, without either the prior written permission of the Publisher, or authorization through payment of the appropriate per-copy fee to the Copyright Clearance Center, 222 Rosewood Drive, Danvers, MA 01923, (978) 750-8400, fax (978) 646-8600, or on the web at www.copyright.com. Requests to the Publisher for permission should be addressed to the Permissions Department, John Wiley & Sons, Inc., 111 River Street, Hoboken, NJ 07030, (201) 748-6011, fax (201) 748-6008, or online at www.wiley.com/go/permissions.

Limit of Liability/Disclaimer of Warranty: While the publisher and author have used their best efforts in preparing this book, they make no representations or warranties with respect to the accuracy or completeness of the contents of this book and specifically disclaim any implied warranties of merchantability or fitness for a particular purpose. No warranty may be created or extended by sales representatives or written sales materials. The advice and strategies contained herein may not be suitable for your situation. You should consult with a professional where appropriate. Neither the publisher nor the author shall be liable for damages arising herefrom.

For general information about our other products and services, please contact our Customer Care Department within the United States at (800) 762-2974, outside the United States at (317) 572-3993 or fax (317) 572-4002.

Wiley publishes in a variety of print and electronic formats and by print-on-demand. Some material included with standard print versions of this book may not be included in e-books or in print-on-demand. If this book refers to media such as a CD or DVD that is not included in the version you purchased, you may download this material at http://booksupport.wiley.com. For more information about Wiley products, visit www.wiley.com.

Library of Congress Cataloging-in-Publication Data:
Names: Simpson, Scot, author.
Title: Complete book of framing : an illustrated guide for residential construction / Scot Simpson.
Description: Second edition—updated and expanded. | New Jersey : John Wiley & Sons, Inc., [2019] | Includes index. |
Identifiers: LCCN 2018054066 (print) | LCCN 2018054893 (ebook) | ISBN 9781119528500 (AdobePDF) | ISBN 9781119528517 (ePub) | ISBN 9781119528524 (paper)
Subjects: LCSH: Framing (Building) | Wooden-frame buildings—Design and construction. | House framing.
Classification: LCC TH2301 (ebook) | LCC TH2301 .S483 2019 (print) | DDC 694/.2—dc23
LC record available at https://lccn.loc.gov/2018054066

Printed in the United States of America
SKY10047351_051223

TABLE OF CONTENTS

ABOUT THE AUTHOR

Scot Simpson has recently retired from a lifetime of framing houses, schools, and commercial buildings for 41 years. He owned a construction firm for 36 years. His firm, S.S. Framing, Inc., was based in Edmonds, WA. He developed and refined the methods in this book and used them to train his crews. Scot is the author of two other construction books and many articles for construction magazines, such as *Fine Homebuilding* and the *Journal of Light Construction*. He developed and hosted the video "Resisting the Forces of Earthquakes" with the Earthquake Engineering Research Institute and the International Conference of Building Officials.

Scot is a member of the International Code Council (ICC), the Construction Specifications Institute (CSI), and the Associated General Contractors of America (AGC), and was 2006 Chairman of the ABC Framers Council. He has presented training and seminars for the National Association of Homebuilders, the American Forest and Paper Association, and the International Conference of Building Officials, among others, in the U.S., Japan, Korea, the Czech Republic, Bulgaria, Spain, Greece, and Mexico.

Scot holds an MBA from Kent State University, as well as a BA and technical certificates in carpentry instruction, lumber grading, and industrial first aid.

ACKNOWLEDGMENTS

The author appreciates and would like to acknowledge the following individuals and organizations whose efforts and documents have provided content for this book:

Allan R. Simpson, Jr.; Dr. Alan Kelley; Lara Simpson, Bruce Simpson; Mars Simpson; Casey Miller; Dave Neiger; Jeff Harding; John E. Farrier APA, the Engineered Wood Association; The Association of Mechanical Engineers (ASME); Digital Canal Corporation; iLevel, a Weyerhaeuser Business, Boise, Idaho; The International Code Council (ICC); The Mason Contractors Association of America (MCAA); Simpson Lumber Company; The Simpson Strong-Tie Company; The Truss Plate Institute; the U.S. Geological Survey National Seismic Hazard Mapping Project; the Western Wood Products Association (WWPA); and Premier Building Systems.

INTRODUCTION

I was a framing contractor for 36 years. I've spent most of my career as a lead framer, directing my framing crews and training workers to become framers. In my teaching, I found that much of the information I needed was not available in a good book, so I wrote one, *Framing & Rough Carpentry*. As I started spending more of my time training and working with lead framers, I again looked for a good, easy-to-understand reference. I didn't find what I needed, so I wrote another book, *Advanced Framing Methods*, that provides all the information a framer needs to move up to the next level—becoming a lead framer. *The Complete Book of Framing* is the combination of those two books, updated with full-color illustrations and photographs, plus additional information—all presented in what I've come to think of as a "framer-friendly" format.

Now, as a retired framer, I realize how much I abused my body during a lifetime of framing. While updating this book I added a section on "Healthy Framing"—what you need to know about how framing affects your body; and what you can do to minimize those effects.

If you're a novice with no framing experience, you'll see the basics of framing shown in a simple, step-by-step style that makes it easy to learn. Where possible, I included both photographs and drawings for each step—for quick and complete learning. The advanced information will be more difficult for a novice to understand, but getting a good feel for the framing basics that come before it will help. The more advanced tasks are also explained with photos and clear drawings.

If you're already an experienced framer, the book gives you some unique tools that you won't find anywhere else. For example, after struggling with rafters and rake walls for years, I developed a diagonal percent system that makes it easier. I use this for finding rafter lengths and rake wall stud heights. The book also explains all the "classic" methods for doing these tasks, but once you try the diagonal percent system, I doubt you'll go back to the old methods. Another example of the book's unique style of presentation is the layout language, which I developed for my first book.

If you're a lead framer, all the basic framing steps presented are important for reference and to help you teach and train crews. Most valuable, however, will be the guidance on managing a framing crew. Once you become a lead framer, your productivity is defined by the productivity of your crew. You'll need to think about the information they need and how to teach and manage them most effectively. Chapter 14 of this book is like a mini framer management course.

Contents

Chapter One

INTRODUCTION TO FRAMING

The trade of wood framing comprises the rough carpentry skills needed to produce the "skeleton" of a building and its first layer of "skin." The skeleton consists of the structural lumber forming the floors, walls, and roof. The skin consists of the lumber that encloses the skeleton and provides a surface for subsequent layers of protective and decorative finish materials.

This chapter is an illustrated review of a framer's most basic tools, materials, and terminology. This basic information is often not even taught on the job site, so if you don't know it when you arrive for work, you will have to play a guessing game or ask a lot of questions.

The detailed illustrations serve as a handy reference and help to reduce confusion when different words are used for the same item. Confusion can arise when framers move from job site to job site and work with different people. For example, bottom plates are often known as *sole plates*, backers as

partitions, and trimmers as *jack studs*. But it doesn't matter what they are called as long as you know what they are. There is also a more detailed list of framing terms with definitions at the back of the book.

The suggested organization for a framing tool truck presented in this chapter is just an example of how a truck might be set up for tool storage. Its purpose is, once again, to reduce confusion and make the job easier. It is amazing how much time can be spent looking for tools and nails if they aren't put where you expect them to be.

Framing Terms

Bearing Walls

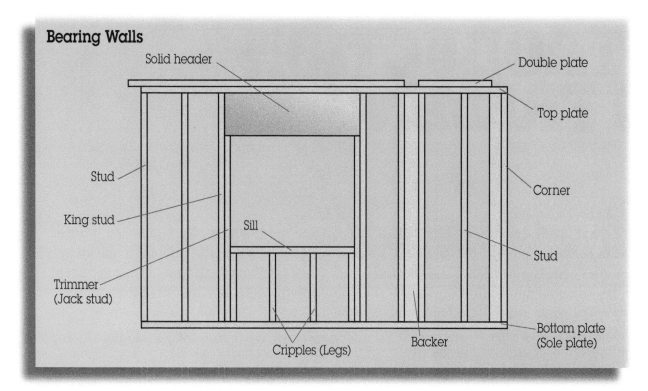

Bearing walls support the main weight of an upper portion of a building, such as a ceiling, floor, or roof. Nonbearing walls provide little or no support to those upper portions. Remove nonbearing walls, and the upper portions will stand; remove bearing walls, and the upper portions will fall.

Nonbearing Walls

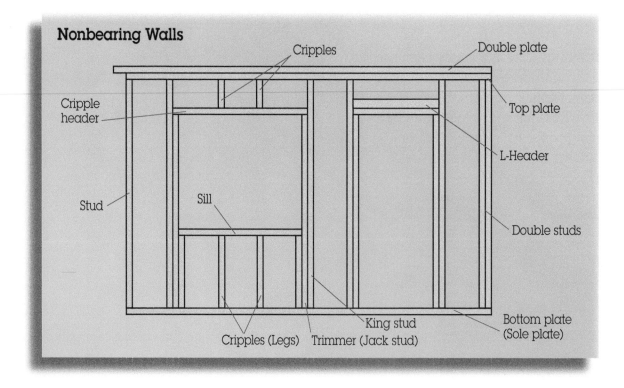

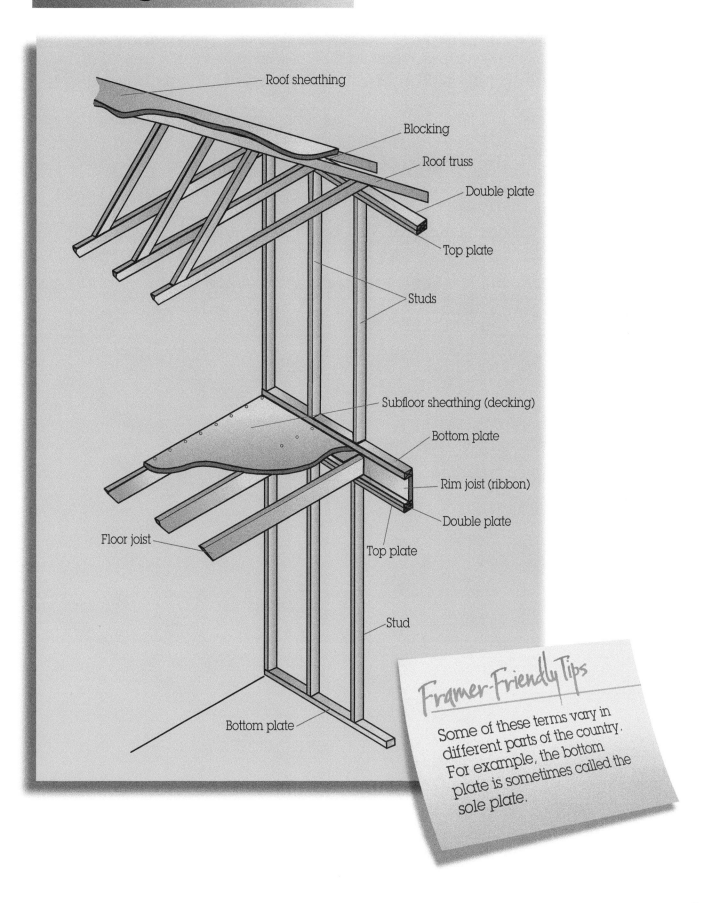

Roof sheathing

Blocking

Roof truss

Double plate

Top plate

Studs

Subfloor sheathing (decking)

Bottom plate

Rim joist (ribbon)

Double plate

Top plate

Floor joist

Stud

Bottom plate

Framer-Friendly Tips

Some of these terms vary in different parts of the country. For example, the bottom plate is sometimes called the sole plate.

Framing Lumber

Lumber is sized in "nominal," as opposed to "actual," dimensions. A nominal dimension rounds off the actual dimension to the next highest whole number. For example, a piece of lumber that actually measures 1-½" × 3-½" is rounded off to the nominal 2" × 4".

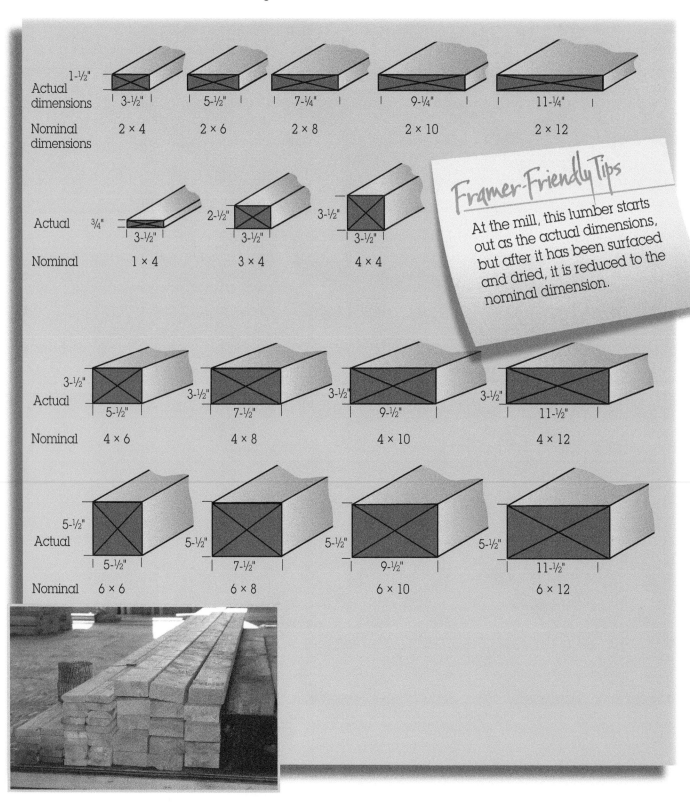

Actual dimensions
| 1-½" | | | | |
| 3-½" | 5-½" | 7-¼" | 9-¼" | 11-¼" |

Nominal dimensions
| 2 × 4 | 2 × 6 | 2 × 8 | 2 × 10 | 2 × 12 |

Actual
| ¾" | 2-½" | 3-½" |
| 3-½" | 3-½" | 3-½" |

Nominal
| 1 × 4 | 3 × 4 | 4 × 4 |

Framer-Friendly Tips

At the mill, this lumber starts out as the actual dimensions, but after it has been surfaced and dried, it is reduced to the nominal dimension.

Actual
| 3-½" | 3-½" | 3-½" | 3-½" |
| 5-½" | 7-½" | 9-½" | 11-½" |

Nominal
| 4 × 6 | 4 × 8 | 4 × 10 | 4 × 12 |

Actual
| 5-½" | 5-½" | 5-½" | 5-½" |
| 5-½" | 7-½" | 9-½" | 11-½" |

Nominal
| 6 × 6 | 6 × 8 | 6 × 10 | 6 × 12 |

Framing Sheathing

Engineered Panel Products

Sheathing comes in 4' × 8' sheets. The thicknesses most commonly used in framing are ½", ⅝", and ¾".

Framer-Friendly Tips

There are substitutions for these standard sizes. For example, $^7/_{16}$" and $^{15}/_{32}$" are common substitutions for ½".

½" sheet
⅝" sheet
¾" sheet

The engineered panel products on this wall provide the strength needed for the high ceiling of this elementary school.

T&G Wood Structural Panels
(tongue and groove)

½" — GWB
⅝" — GWB

GWB = Gypsum wallboard (also called drywall or sheetrock). The most common thicknesses are ½" and ⅝".

Although not as common in house framing, gypsum wallboard can be used on exterior walls, such as for apartments and condos, and commercial buildings for fire protection.

Dens Glass® gypsum sheathing is a brand that has fiberglass mat, which provides mold and moisture resistance and is gold in color.

Engineered Wood Products

Engineered wood products are becoming more and more a part of our everyday framing. The strengths of these different products vary. Whenever you use engineered wood, it is important that you understand the qualities of the specific product you are planning to use, as well as structural considerations and any restrictions on cutting and installation.

Engineered wood products can be divided into two categories: **engineered panel products** and **engineered lumber products**. Engineered panel products include plywood, oriented strand board (OSB), waferboard, composite, and structural particleboard. Engineered lumber products include I-joists, glu-lam beams, LVLs (laminated veneer lumber), PSLs (parallel strand lumber), LSLs (laminated strand lumber), OSL (oriented strand lumber), and CLT (cross-laminated timber.)

Engineered wood products have structural qualities different than those of traditional wood, so they must be used within the specification set by the manufacturer. When these products are specified on the plans, the architect or engineer who specified them will have checked with the structural engineer to ensure proper use.

Engineered panel products have been around for years and are treated in a

I-joists are engineered lumber products that provide consistency and fewer floor squeaks.

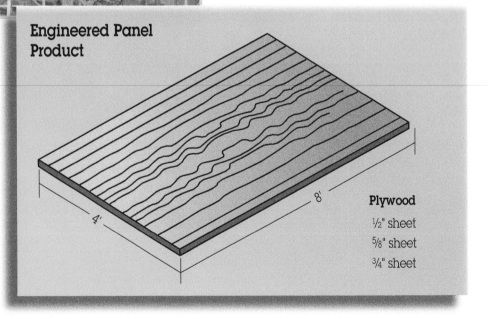

Engineered Panel Product

Plywood

½" sheet

⅝" sheet

¾" sheet

8'

4'

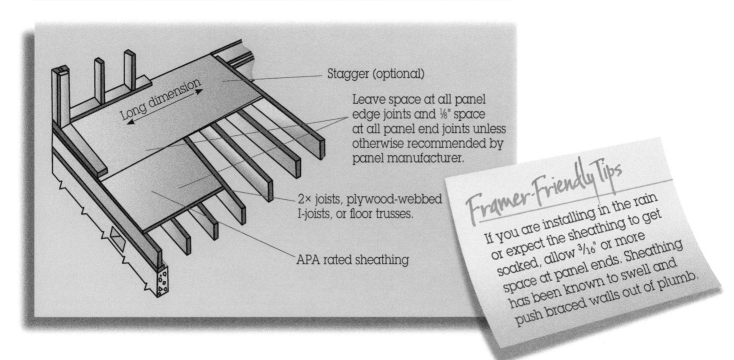

Stagger (optional)

Leave space at all panel edge joints and ⅛" space at all panel end joints unless otherwise recommended by panel manufacturer.

Long dimension

2× joists, plywood-webbed I-joists, or floor trusses.

APA rated sheathing

Framer-Friendly Tips

If you are installing in the rain or expect the sheathing to get soaked, allow ³⁄₁₆" or more space at panel ends. Sheathing has been known to swell and push braced walls out of plumb.

manner similar to engineered wood products. The 4' × 8' typical sheets are strongest in the direction of the grain. For floors and roofs, these sheets should be laid perpendicular to the direction of the supporting members. The strength of the panels comes from the panel cantilevering over the supports—so each piece should be at least as long as two support members.

Glu-lam beams, LVLs, PSLs, and LSLs can be cut to length, but should not be drilled or notched without checking with manufacturers' specifications.

I-joists are becoming more widely used. Although the Engineered Wood Association has a standard for I-joists, not all I-joists manufacturers subscribe to that standard. Consequently, it is important to follow the manufacturer's instructions whenever using I-joists. Installation instructions are usually delivered with the load for each job. The illustration shows some of the typical instructions.

Certain features are common among all I-joists. Rim and blocking may be of I-joist or solid rim board. Typical widths are 9-½", 11-⅞", 14", 16", and 20".

Web stiffeners are used to add to the strength at bearing points. If the bearing point is at the bottom flange, then the web stiffener, which is the thickness of the flange on one side of the web, is held tight to the bottom. There should be at least a ⅛" space between the top flange and the web stiffener. If the bearing point is at the top flange, then the web stiffener is held tight to the top with at least ⅛" between the bottom flange and the web stiffener.

Framer-Friendly Tips

Special details are needed when attaching heavy weight—such as blocking for hanging cast iron pipes, to the bottom of I-joists—to prevent the bottom flange from breaking the glue that attaches it to the web.

Engineered Wood Products (continued)

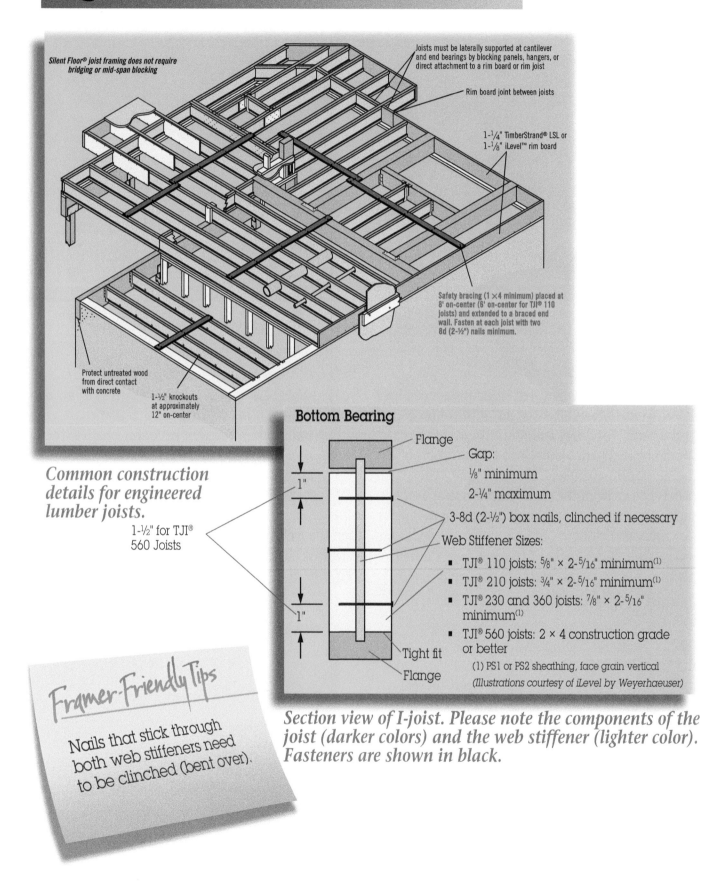

Silent Floor® joist framing does not require bridging or mid-span blocking

Joists must be laterally supported at cantilever and end bearings by blocking panels, hangers, or direct attachment to a rim board or rim joist

Rim board joint between joists

1-¼" TimberStrand® LSL or 1-⅛" iLevel™ rim board

Safety bracing (1 × 4 minimum) placed at 8' on-center (6' on-center for TJI® 110 joists) and extended to a braced end wall. Fasten at each joist with two 8d (2-½") nails minimum.

Protect untreated wood from direct contact with concrete

1-½" knockouts at approximately 12" on-center

Common construction details for engineered lumber joists.

1-½" for TJI® 560 Joists

Bottom Bearing

Flange

1"

1"

Gap:
⅛" minimum
2-¼" maximum

3-8d (2-½") box nails, clinched if necessary

Web Stiffener Sizes:

- TJI® 110 joists: ⅝" × 2-5/16" minimum(1)
- TJI® 210 joists: ¾" × 2-5/16" minimum(1)
- TJI® 230 and 360 joists: ⅞" × 2-5/16" minimum(1)
- TJI® 560 joists: 2 × 4 construction grade or better

Tight fit

Flange

(1) PS1 or PS2 sheathing, face grain vertical

(Illustrations courtesy of iLevel by Weyerhaeuser)

Framer-Friendly Tips

Nails that stick through both web stiffeners need to be clinched (bent over).

Section view of I-joist. Please note the components of the joist (darker colors) and the web stiffener (lighter color). Fasteners are shown in black.

10

Engineered Wood Products (continued)

Squash blocks are pieces of lumber installed alongside TJIs at points of heavy loading. They prevent the weight from crushing the TJI. They are typically dimensional lumber like 2 × 4s or 2 × 6s. They should be cut 1/16" longer than the I-joist to take the load off the I-joists.

I-joist hardware, such as hangers, is usually delivered with the I-joist package. However, standard I-joist hardware can be purchased separately.

I-joists typically require a 1-¾" bearing. You can cut the end of an I-joist as long as it is not cut beyond a line straight up from the end of the bearing. However, no cuts should extend beyond a vertical line drawn from the end of the bearing point.

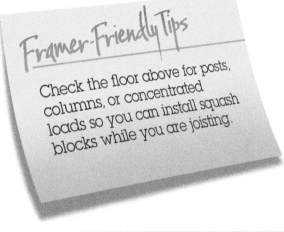

Framer-Friendly Tips

Check the floor above for posts, columns, or concentrated loads so you can install squash blocks while you are joisting.

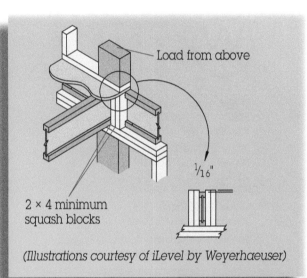

Load from above

2 × 4 minimum squash blocks

1/16"

(Illustrations courtesy of iLevel by Weyerhaeuser)

Squash blocks should be 1/16" greater than the I-joist height.

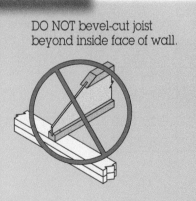

DO NOT bevel-cut joist beyond inside face of wall.

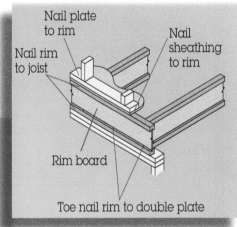

Nail plate to rim

Nail rim to joist

Nail sheathing to rim

Rim board

Toe nail rim to double plate

Lumber and Wood Structural Panel Grade Stamps

Lumber and wood structural panels are graded for strength and different uses. Each piece of lumber is stamped for identification before it is shipped.

Architects specify grades of lumber and wood structural panels for various purposes, and framers need to make sure the right wood is used.

Wood Structural Panels

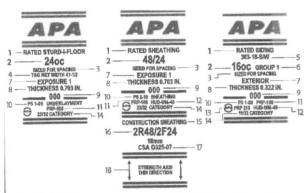

1. Panel grade
2. Span rating
3. Sized less than full length/width
4. Tongue-and-groove
5. Siding face grade
6. Species group number
7. Bond classification
8. Decimal thickness designation (this value is generally at or near the lower tolerance specified in PS 1 or PS 2)
9. Mill number
10. Product standard
11. APA's performance rated panel standard
12. HUD recognition
13. Referenced product standard
14. Performance category
15. Panel grade, Canadian standard
16. Panel mark—Rating and end-use designation per the Canadian standard
17. Canadian performance rated panel standard
18. Panel face grain orientation indicator

a. **WWPA certification mark** certifies Association Quality standards and is a registered trademark.

b. **Mill identification** Firm name, brand, or assigned mill number. WWPA can be contacted to identify an individual mill whenever necessary.

c. **Grade designation** Grade name, number, or abbreviation.

d. **Species identification** Indicates species by individual species or species combination.

e. **Condition of seasoning** Indicates condition of seasoning at time of surfacing:

MC-15 – KD-15: 15% maximum moisture content

S-DRY – KD: 19% maximum moisture content

S-GRN: over 19% moisture content (unseasoned)

Lumber

Framing	Grade	Use
Light framing 2 × 2 thru 4 × 4	Construction Standard and better Utility	Plates Sills Studs
Stud 2 × 2 thru 4 × 18	Stud	Studs Cripples
Structural framing	Select structural No. 1 and BTR No. 1 No. 2 No. 3	Studs Plates Joists Rafters Headers Posts Beams

Framer-Friendly Tips

The specification on your plans should tell you the grade you need to use.

Images of grade stamps courtesy of APA, The Engineered Wood Association and WWPA, the Western Wood Products Assocoiation

Framing Nails

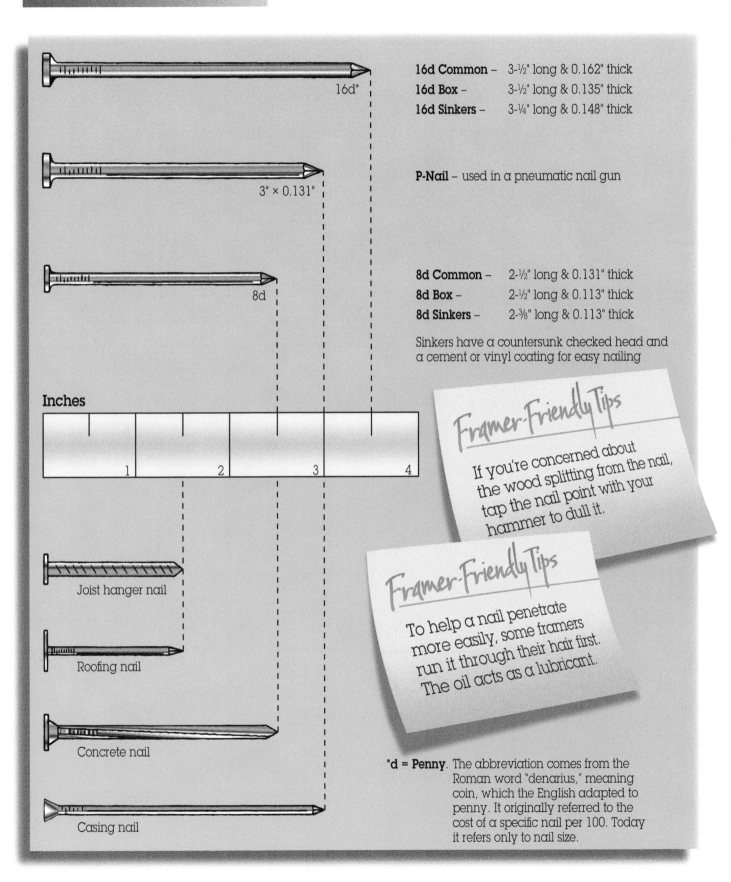

16d*

3" × 0.131"

8d

16d Common –   3-½" long & 0.162" thick
16d Box –   3-½" long & 0.135" thick
16d Sinkers –   3-¼" long & 0.148" thick

P-Nail – used in a pneumatic nail gun

8d Common –   2-½" long & 0.131" thick
8d Box –   2-½" long & 0.113" thick
8d Sinkers –   2-⅜" long & 0.113" thick

Sinkers have a countersunk checked head and a cement or vinyl coating for easy nailing

Inches

1	2	3	4

Joist hanger nail

Roofing nail

Concrete nail

Casing nail

Framer-Friendly Tips

If you're concerned about the wood splitting from the nail, tap the nail point with your hammer to dull it.

Framer-Friendly Tips

To help a nail penetrate more easily, some framers run it through their hair first. The oil acts as a lubricant.

*d = Penny. The abbreviation comes from the Roman word "denarius," meaning coin, which the English adapted to penny. It originally referred to the cost of a specific nail per 100. Today it refers only to nail size.

Framing Tools

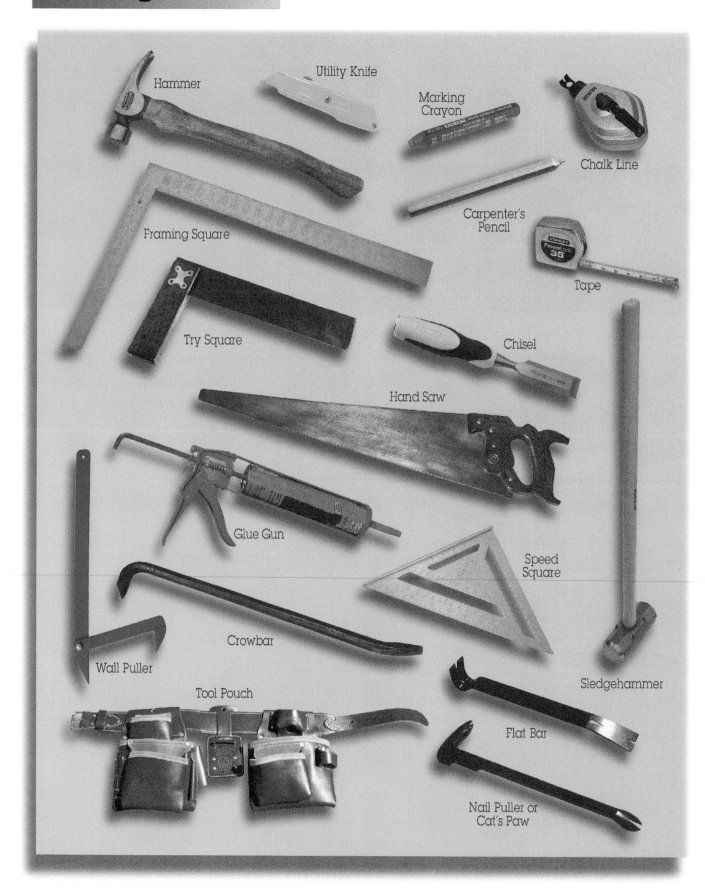

Hammer

Utility Knife

Marking Crayon

Chalk Line

Framing Square

Carpenter's Pencil

Tape

Try Square

Chisel

Hand Saw

Glue Gun

Speed Square

Wall Puller

Crowbar

Sledgehammer

Tool Pouch

Flat Bar

Nail Puller or Cat's Paw

Framing Tools (continued)

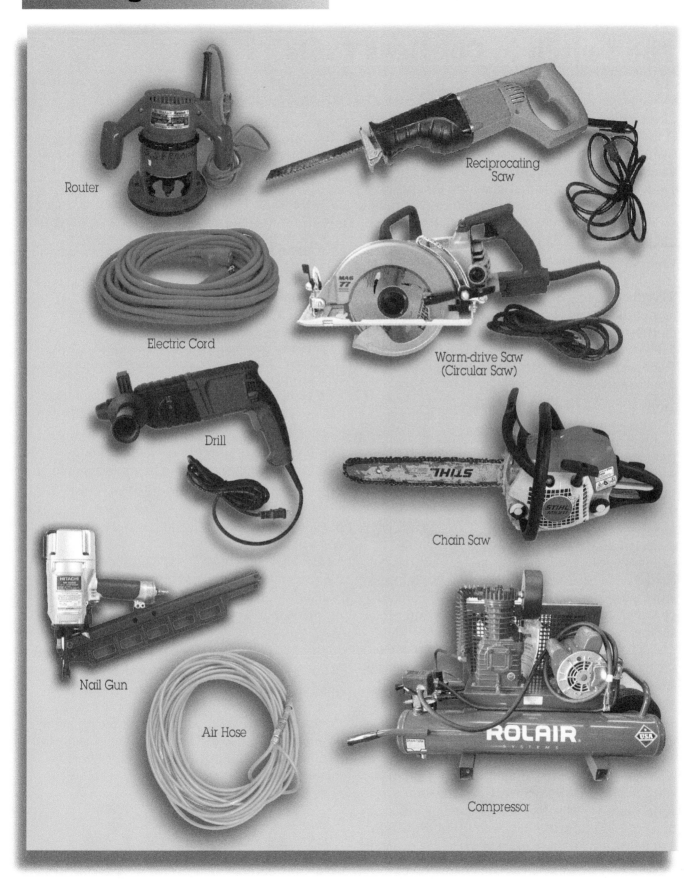

Router

Reciprocating Saw

Electric Cord

Worm-drive Saw (Circular Saw)

Drill

Chain Saw

Nail Gun

Air Hose

Compressor

The Switch to Cordless Tools

It's inevitable at some point that future framers will cut the cord and hose completely. The awkwardness of the cord, and the safety issues with having cords laying around, cause difficulties. With electric cords there is also the problem of rain and keeping the connections dry so that they don't trip the circuit breaker. There is also the time and difficulty of rolling up, out, and storing the cords. All in all, cords are a real pain; however, they do provide a lot of power.

The cord has already been cut for some electric tools, and manufacturers are coming out with more powerful battery tools all the time. Battery tools do have their own set of problems. There must be a plan for recharging and enough batteries to make sure that you don't lose power. There is also the reality that batteries from different manufacturers are not interchangeable.

Cutting the cord requires purchase of battery tools. The decision on the best tool is complicated. Different manufacturer's batteries are not interchangeable; however, similar-voltage batteries of different tools by the same manufacturer typically are interchangeable. It is efficient to have tools of the

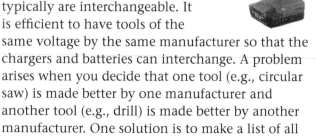

same voltage by the same manufacturer so that the chargers and batteries can interchange. A problem arises when you decide that one tool (e.g., circular saw) is made better by one manufacturer and another tool (e.g., drill) is made better by another manufacturer. One solution is to make a list of all the tools you need and then summarize the benefits and disadvantages of each of the tools from each manufacturer.

To do a brief analysis, it is important to have an understanding of variables that make a battery tool good. To start with, it is important to understand the basics of electricity measurements. Watts (power) is equal to amps (current) times volts (pressure). Watts = amps × volts. Torque is the force produced to turn an object. Torque differs from power or watts in that it depends on the makeup of the tool, while watts indicates the electric power delivered to the tool.

Measurements for Analyzing Tools

1. **Watts:** power = amps × volts
2. **Amps:** the amount of current (electrons flowing)
3. **Volts:** the potential difference = pressure to move the electrons
4. **Torque:** force produced to move an object
5. **HP: H**orse **P**ower—a unit of measurement of power
6. **RPM: R**otation **P**er **M**inute—how fast a tool turns
7. **IPM: I**mpacts **P**er **M**inute—impact wrenches
8. **BPM: B**lows **P**er **M**inute
9. **SPM: S**trokes **P**er **M**inute—reciprocating saws
10. **UWO: U**nits **W**atts **O**ut—power—max speed and torque
11. **VSR: V**ariable **S**peed **R**eversible
12. **AH: A**mp **H**our—battery capacity—one amp of current for one hour
13. **Ft-lbs:** the torque created by one pound of force acting at a perpendicular distance of one foot from a pivot point
14. **In-lbs:** the torque created by one pound of force acting at a perpendicular distance of one inch from a pivot point

The amount of time a battery will last before needing a recharge is the amp hours and should be listed in the specifications for the tool; however, that is not always the case.

Brushless technology is used on many new tools and will probably take over for brushed tools in the future. Brushless eliminates the brushes touching the commutator in the motors and therefore reduces heat and friction, which reduce the power output. Brushless tools are also lighter.

There are other factors that you will want to consider in evaluating battery tools, like the weight of the tool, if it has ergonomic padded hand grips, rafter hooks, electric brakes for circular saws, and variable speed for drills and reciprocating saws. Compare foot- or inch-pounds of torque for hammer drills and impact wrenches, maximum capacity for drill hole sizes, and many other advantages that manufacturers are always coming up with.

Framing Tool Truck

Typical Layout for a 14' Step Van

If you're a professional framer, organizing your tools helps keep them in good condition and helps you find them when you need them—saving valuable time on the job.

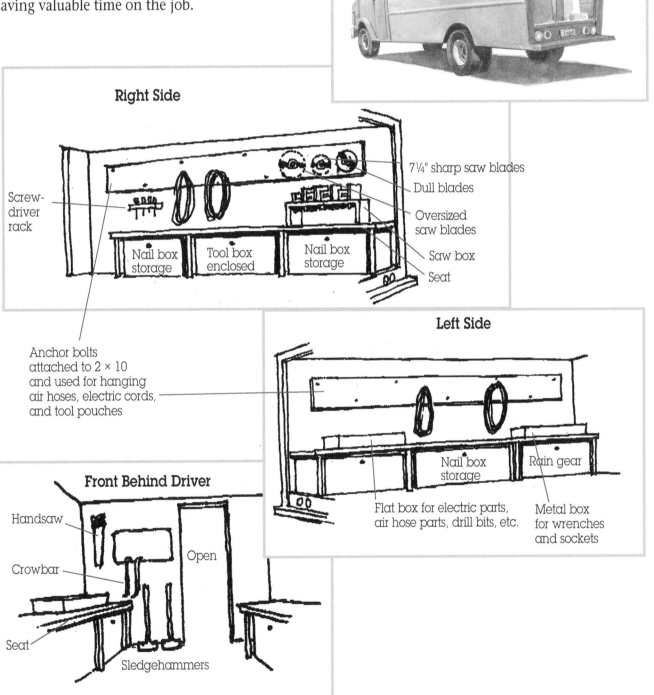

Right Side

Screw-driver rack

7¼" sharp saw blades

Dull blades

Oversized saw blades

Saw box

Seat

Nail box storage

Tool box enclosed

Nail box storage

Anchor bolts attached to 2 × 10 and used for hanging air hoses, electric cords, and tool pouches

Left Side

Nail box storage

Rain gear

Flat box for electric parts, air hose parts, drill bits, etc.

Metal box for wrenches and sockets

Front Behind Driver

Handsaw

Crowbar

Seat

Open

Sledgehammers

Cutting Lumber

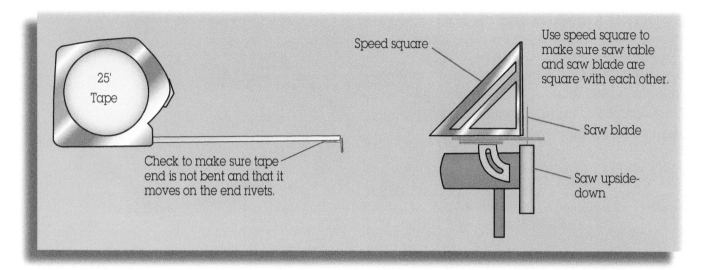

25' Tape

Check to make sure tape end is not bent and that it moves on the end rivets.

Speed square

Use speed square to make sure saw table and saw blade are square with each other.

Saw blade

Saw upside-down

Accuracy in measuring, marking, and cutting lumber is a very important framing skill to master. Periodic checks should be made of the condition of tape measures and the squareness of saw tables and blades.

A typical saw blade removes a channel of wood approximately ⅛" wide, called a *kerf*. This must be taken into consideration when you make a cut.

Suppose you want to cut a board 25" long. Measure and make a mark at 25", then square a line through the mark with a square. The *work piece*— the 25" piece you want to use—will be to the left of the line; the *waste piece* will be to the right. Guide your saw along the right edge of the line so the kerf is made

Framer-Friendly Tips

The tip of the tape is made to move in and out the thickness of the metal tip. This allows for the same measurement whether you butt the tip or hook the tape.

in the waste piece. If your cut is perfectly made, the work piece will be left showing exactly half the width of your pencil line, and will measure exactly 25". Thus, the old carpenter's saying: "Leave the line."

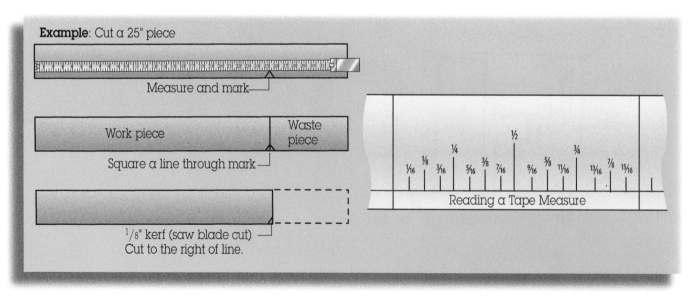

Example: Cut a 25" piece

Measure and mark

Work piece | Waste piece

Square a line through mark

⅛" kerf (saw blade cut) — Cut to the right of line.

Reading a Tape Measure

¹⁄₁₆ ⅛ ³⁄₁₆ ¼ ⁵⁄₁₆ ⅜ ⁷⁄₁₆ ½ ⁹⁄₁₆ ⅝ ¹¹⁄₁₆ ¾ ¹³⁄₁₆ ⅞ ¹⁵⁄₁₆

Protecting Lumber from Decay

Moisture and warmth will promote decay of most woods. To prevent decay, naturally durable woods or preservative-treated wood must be used when the wood is exposed to moisture.

Decay-resistant woods include redwood, cedar, black locust, and black walnut. Preservative-treated wood is treated according to certain industrial specifications. Preservative-treated wood is most commonly used because of its availability.

Preservative-treated or naturally durable woods should be used in the following locations:

1. On concrete foundation walls that are less than 8" from exposed earth.

2. On concrete or masonry slabs that are in direct contact with earth.

3. Where wood is attached directly to the interior of exterior masonry or concrete walls below grade.

4. For floor joists if they are closer than 18" to the exposed ground.

5. For floor girders if they are closer than 12" to the exposed ground.

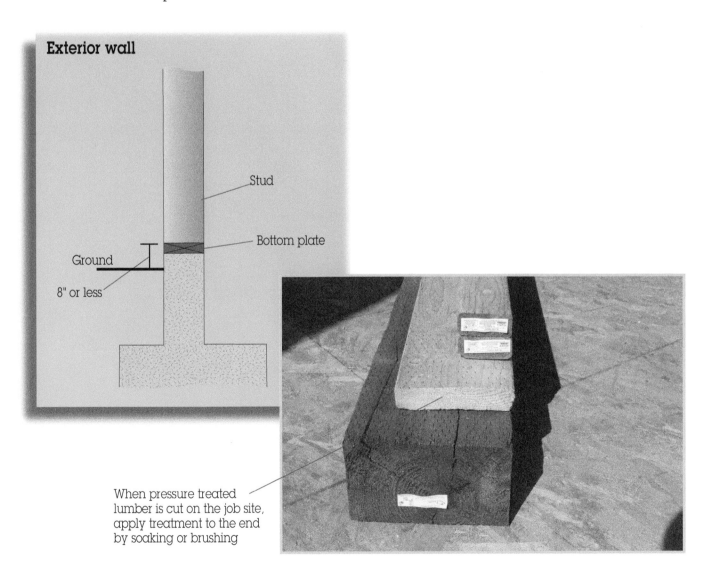

Exterior wall

Stud

Bottom plate

Ground

8" or less

When pressure treated lumber is cut on the job site, apply treatment to the end by soaking or brushing

Protecting Lumber from Decay (continued)

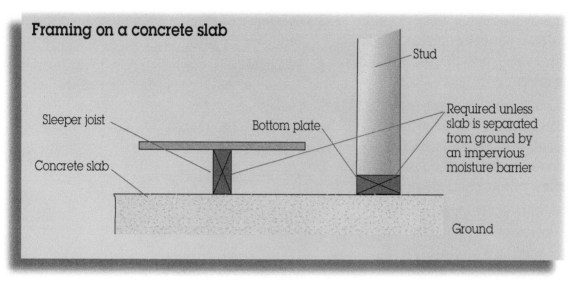

Framing on a concrete slab

Stud

Sleeper joist

Bottom plate

Required unless slab is separated from ground by an impervious moisture barrier

Concrete slab

Ground

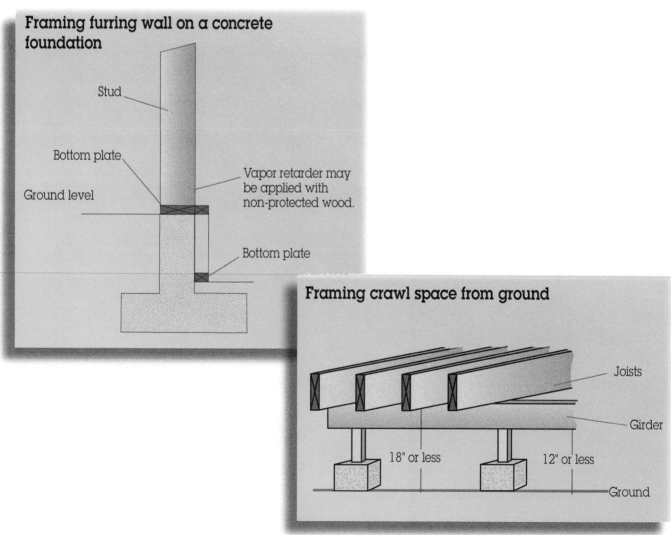

Framing furring wall on a concrete foundation

Stud

Bottom plate

Ground level

Vapor retarder may be applied with non-protected wood.

Bottom plate

Framing crawl space from ground

Joists

Girder

18" or less

12" or less

Ground

Preservative Treated Wood

The treating of wood in recent years has gone through some major changes. The most important thing to know is that there are different types of preservative treatment and that some of the treatments require specially coated fasteners to prevent corrosion.

A little history will help in understanding. For years the predominant chemical for preserving dimension lumber had been chromated copper arsenate (CCA). However, health concerns arose because of the arsenic content in CCA, and in 2004 the Environmental Protection Agency (EPA) required labels on CCA, which had the effect of disallowing the use of CCA-treated wood for most residential uses.

The first commonly used substitutes were copper azole (CA) and alkaline copper quaternary (ACQ). These eliminated the arsenic but created a different problem because they were corrosive to steel fasteners. To solve this problem, hardware manufacturers began making their common fasteners with a galvanized coating. For example, if you see a Simpson Strong-Tie hardware labeled Z-max you know it has been coated so that it can be used with CA and ACQ. Steel nails also had to be coated when used with lumber treated with CA or ACQ. Typically they are galvanized. Stainless steel is a better substitute for hardware and nails because it is less corrosive, but it is expensive.

Sodium Borate (SBX) preservative treatment is another substitute for CCA that does not have the problem of causing corrosion of steel fasteners, however it will wash out of the lumber with liquid exposure. It is specified for use above ground and continuously protected from liquid water.

New products are continually being developed. Carbon-based compounds are among these and could prove to be less corrosive and natural in color.

The 2018 IBC and IRC code states that preservative treated wood should be in accordance with AWPA U1 and M4 (American Wood Protection Association Use Category System) for the species, product, preservative, and end use. The lumber tag attached to the treated wood will give the use category to assist you in making sure you are using the correctly treated wood.

All the different labels and chemicals can be confusing. Most importantly, make sure that you are using the right treatment for the task at hand and that you are using corrosive resistant fasteners where necessary. To check the correct use of treated lumber, read the tag attached to the lumber or ask the lumber supplier. For CA or ACQ treatment, you will need corrosion-resistant fasteners; for SBX or other borate treatments, you will not need corrosion-resistant fasteners. Beyond that, check on the fastener boxes for specifications or ask the lumber or fastener supplier.

Chapter Two
NAILING PATTERNS

Contents

Chapter Two

NAILING PATTERNS

If you are framing every day, the nailing patterns in this chapter will soon become second nature. For the part-time framer, they can serve as a quick reference.

Building codes and generally accepted practices were followed in developing the nailing patterns in this chapter. When the plans call for other nailing patterns, however, be sure to follow them.

You will notice in this chapter that there are different nails specified for the same nailing. There are many different styles of nails. The five most frequently used categories are:

- Common nails
- Box nails
- Sinker nails
- Gun nails
- Positive placement nails

You will see a 3" × 0.131" nail specified frequently. This nail is the most common P-nail, or pneumatic gun nail, used.

The common nails are listed because they are typically specified by building codes. Most of the tests that are done to determine the strength needed use common nails. Box nails and sinker nails are listed because they are easier to nail, and less likely to split the wood. They are also commonly found at nail suppliers. The gun nails are listed because nail guns are used most often. Positive placement nails are made specially for nailing on hardware. They only work in positive placement nail guns.

Please note that common nails are listed with "common" written after the size. If the nail size has "common" after it, you can only use common nails. If it does not, you can use either common, box, or sinker nails.

The *International Residential Code* (IRC) is similar to the *International Building Code* (IBC) except it only covers one- and two-family dwellings. The patterns in this chapter are based on the 2018 IBC, which, in some cases, lists more nail options than the 2018 IRC.

Nail Top Plate to Studs

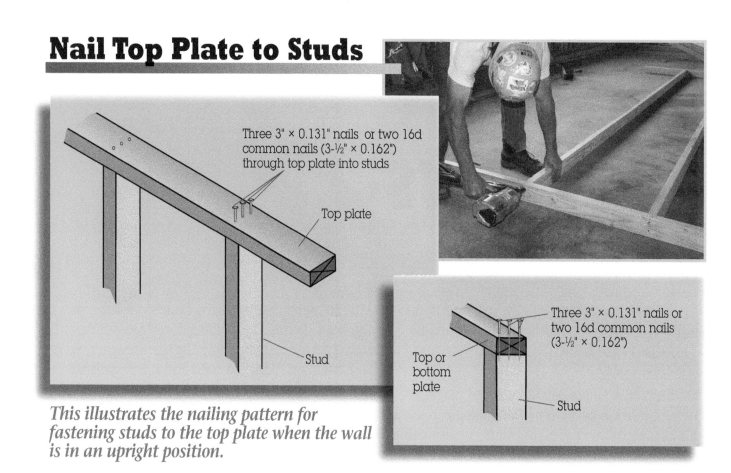

Three 3" × 0.131" nails or two 16d common nails (3-½" × 0.162") through top plate into studs

Top plate

Stud

Three 3" × 0.131" nails or two 16d common nails (3-½" × 0.162")

Top or bottom plate

Stud

This illustrates the nailing pattern for fastening studs to the top plate when the wall is in an upright position.

Nail Bottom Plate to Studs

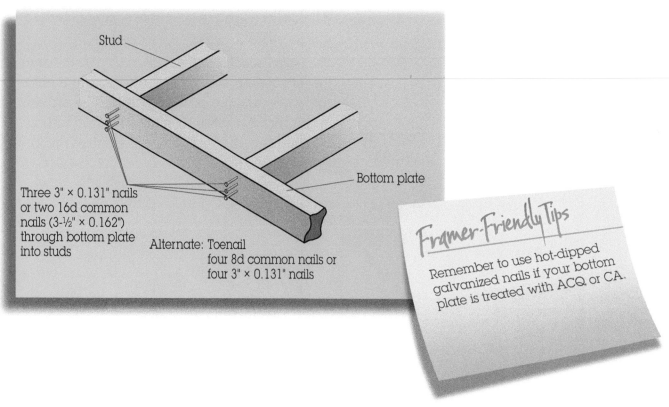

Stud

Bottom plate

Three 3" × 0.131" nails or two 16d common nails (3-½" × 0.162") through bottom plate into studs

Alternate: Toenail four 8d common nails or four 3" × 0.131" nails

Framer-Friendly Tips

Remember to use hot-dipped galvanized nails if your bottom plate is treated with ACQ or CA.

Nail Double Plate to Top Plate

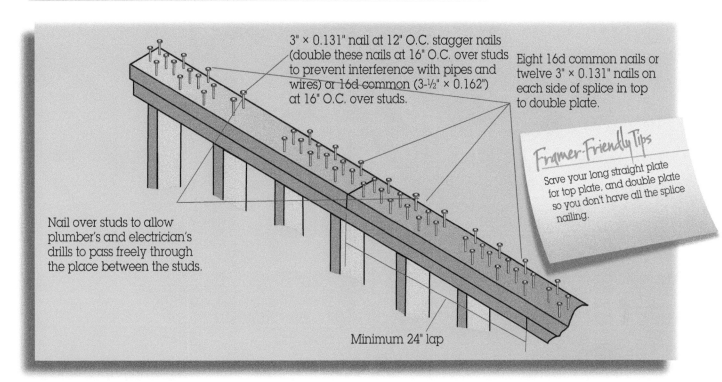

3" × 0.131" nail at 12" O.C. stagger nails (double these nails at 16" O.C. over studs to prevent interference with pipes and wires) or 16d common (3-½" × 0.162") at 16" O.C. over studs.

Eight 16d common nails or twelve 3" × 0.131" nails on each side of splice in top to double plate.

Framer-Friendly Tips

Save your long straight plate for top plate, and double plate so you don't have all the splice nailing.

Nail over studs to allow plumber's and electrician's drills to pass freely through the place between the studs.

Minimum 24" lap

Nail Corner

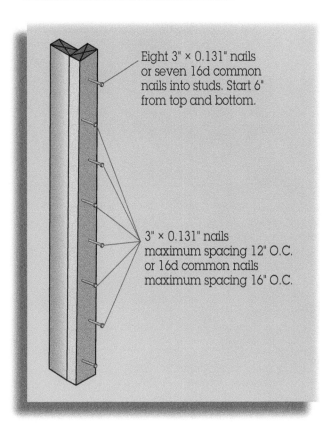

Eight 3" × 0.131" nails or seven 16d common nails into studs. Start 6" from top and bottom.

3" × 0.131" nails maximum spacing 12" O.C. or 16d common nails maximum spacing 16" O.C.

Nail Walls Together or Nail Double Studs

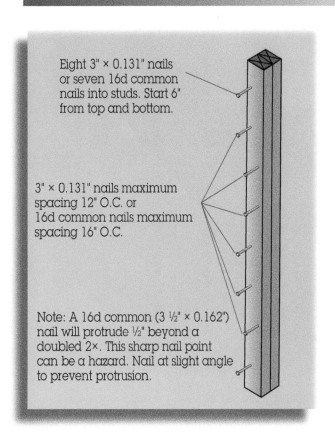

Eight 3" × 0.131" nails or seven 16d common nails into studs. Start 6" from top and bottom.

3" × 0.131" nails maximum spacing 12" O.C. or 16d common nails maximum spacing 16" O.C.

Note: A 16d common (3 ½" × 0.162") nail will protrude ½" beyond a doubled 2×. This sharp nail point can be a hazard. Nail at slight angle to prevent protrusion.

Nail Trimmer to Stud

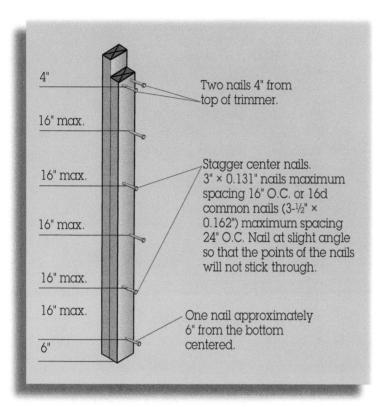

4"

Two nails 4" from top of trimmer.

16" max.

16" max.

Stagger center nails. 3" × 0.131" nails maximum spacing 16" O.C. or 16d common nails (3-½" × 0.162") maximum spacing 24" O.C. Nail at slight angle so that the points of the nails will not stick through.

16" max.

16" max.

16" max.

One nail approximately 6" from the bottom centered.

6"

Concrete Nailing

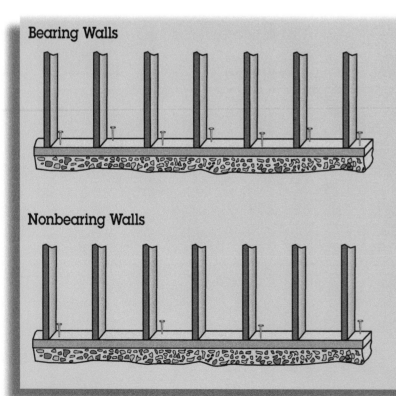

Bearing Walls

Nonbearing Walls

One 2½" concrete nail at 16" O.C., or at every stud.

When anchor bolts are used to secure the bottom plate to the concrete, concrete nails are only needed between the anchor bolts when necessary to straighten the plate or secure the ends of the plate.

One 2½" concrete nail at 32" O.C., or at every other stud.

Framer-Friendly Tips

PATs (powder accuated tools) use a controlled explosion similar to firearms and shoot specially hardened steel nails. They make nailing into concrete easy but noisy.

Nail Bearing and Nonbearing Walls to Floor Perpendicular to Joists

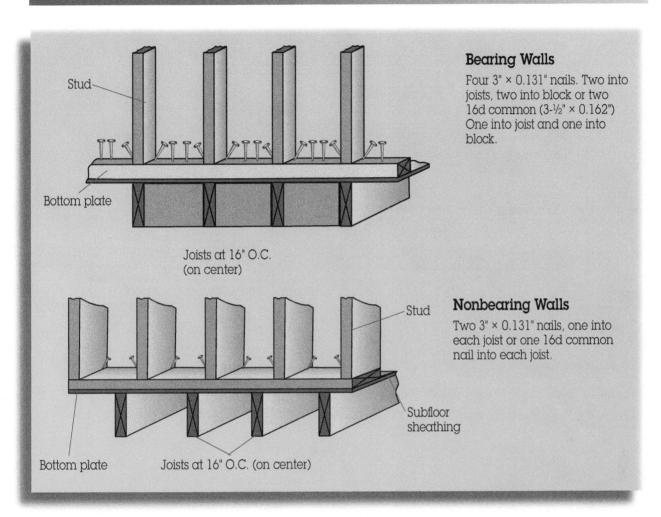

Stud

Bottom plate

Joists at 16" O.C. (on center)

Stud

Subfloor sheathing

Bottom plate

Joists at 16" O.C. (on center)

Bearing Walls

Four 3" × 0.131" nails. Two into joists, two into block or two 16d common (3-½" × 0.162") One into joist and one into block.

Nonbearing Walls

Two 3" × 0.131" nails, one into each joist or one 16d common nail into each joist.

Nail Bearing and Nonbearing Walls to Floor Parallel to Joists

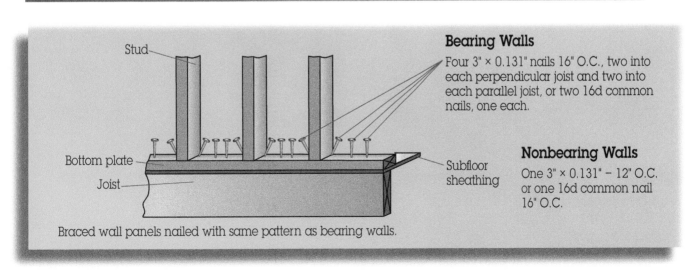

Stud

Bottom plate

Joist

Subfloor sheathing

Bearing Walls

Four 3" × 0.131" nails 16" O.C., two into each perpendicular joist and two into each parallel joist, or two 16d common nails, one each.

Nonbearing Walls

One 3" × 0.131" – 12" O.C. or one 16d common nail 16" O.C.

Braced wall panels nailed with same pattern as bearing walls.

Nail Header to Stud

4×8 or 6×8 Header

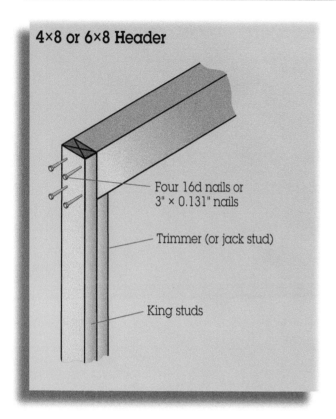

Four 16d nails or
3" × 0.131" nails

Trimmer (or jack stud)

King studs

4×10 or 6×10 Header

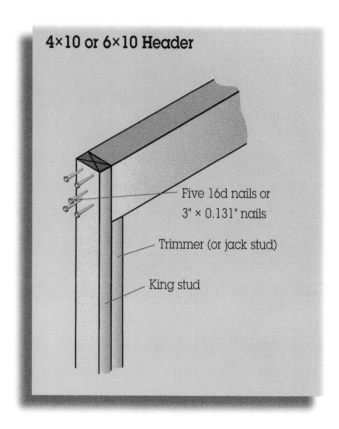

Five 16d nails or
3" × 0.131" nails

Trimmer (or jack stud)

King stud

4×12 or 6×12 Header

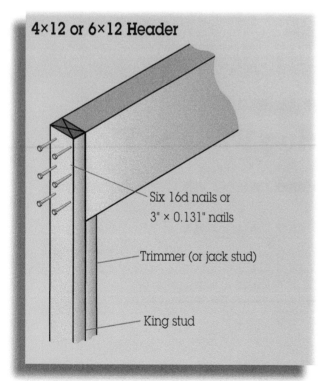

Six 16d nails or
3" × 0.131" nails

Trimmer (or jack stud)

King stud

L - Header

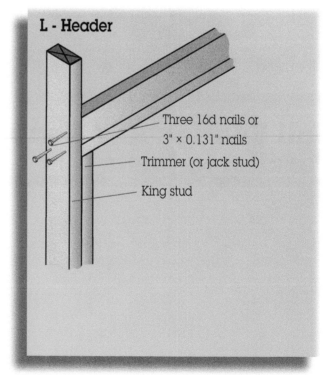

Three 16d nails or
3" × 0.131" nails

Trimmer (or jack stud)

King stud

Headers made up of 2× lumber with ½"
plywood sandwiched between should be
nailed similarly.

Nail Let-in Bracing

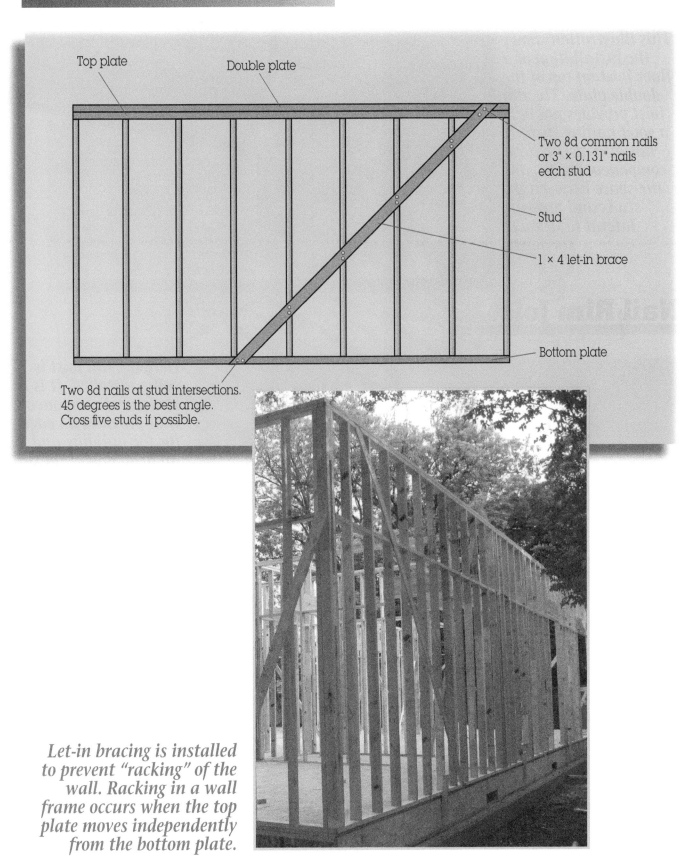

Top plate

Double plate

Two 8d common nails or 3" × 0.131" nails each stud

Stud

1 × 4 let-in brace

Bottom plate

Two 8d nails at stud intersections. 45 degrees is the best angle. Cross five studs if possible.

Let-in bracing is installed to prevent "racking" of the wall. Racking in a wall frame occurs when the top plate moves independently from the bottom plate.

Nail Built-up Girders and Beams

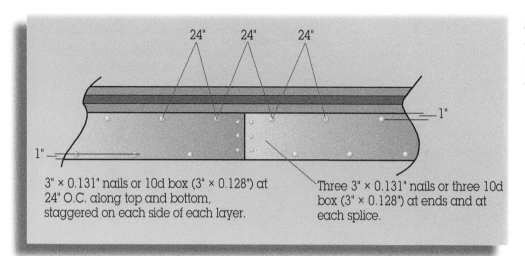

24"   24"   24"

1"

1"

3" × 0.131" nails or 10d box (3" × 0.128") at 24" O.C. along top and bottom, staggered on each side of each layer.

Three 3" × 0.131" nails or three 10d box (3" × 0.128") at ends and at each splice.

Make sure edges are aligned when nailing girders and beams together.

Nail Joist Blocking

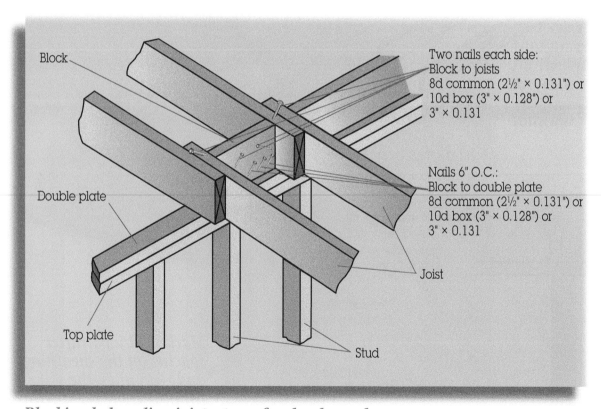

Block

Two nails each side:
Block to joists
8d common (2½" × 0.131") or
10d box (3" × 0.128") or
3" × 0.131

Double plate

Nails 6" O.C.:
Block to double plate
8d common (2½" × 0.131") or
10d box (3" × 0.128") or
3" × 0.131

Joist

Top plate

Stud

Blocking helps align joists, transfers loads, and provides fire protection. The right size block will tighten up the joists without moving them.

Nail Lapping Joists

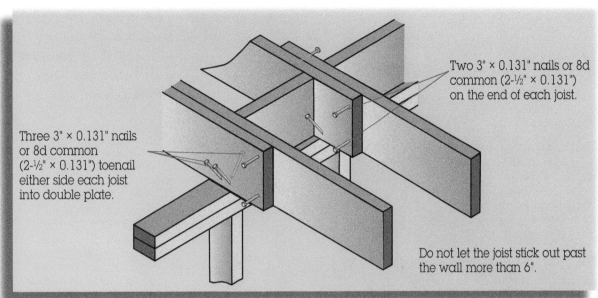

Three 3" × 0.131" nails or 8d common (2-½" × 0.131") toenail either side each joist into double plate.

Two 3" × 0.131" nails or 8d common (2-½" × 0.131") on the end of each joist.

Do not let the joist stick out past the wall more than 6".

Nail Drywall Backing

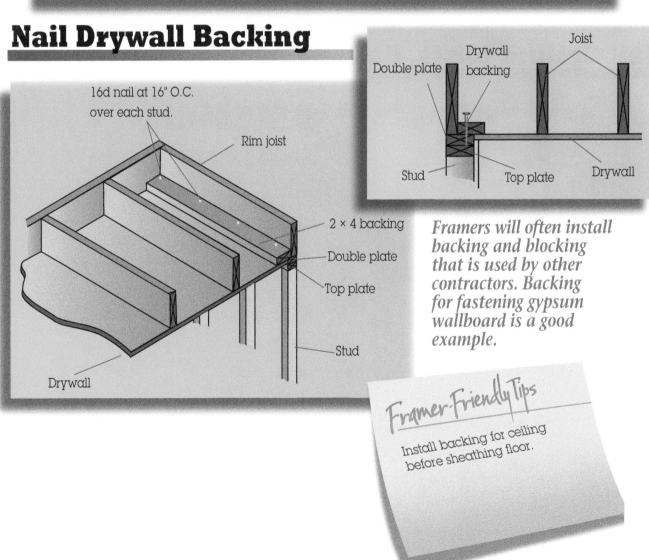

16d nail at 16" O.C. over each stud.

Rim joist

2 × 4 backing

Double plate

Top plate

Stud

Drywall

Double plate

Drywall backing

Joist

Stud

Top plate

Drywall

Framers will often install backing and blocking that is used by other contractors. Backing for fastening gypsum wallboard is a good example.

Framer-Friendly Tips

Install backing for ceiling before sheathing floor.

35

Nail Trusses to Wall

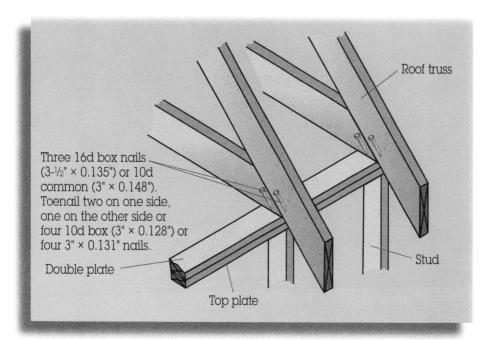

The installation of roof trusses may require the use of a crane or boom truck to lift and position the trusses.

Roof truss

Three 16d box nails (3-½" × 0.135") or 10d common (3" × 0.148"). Toenail two on one side, one on the other side or four 10d box (3" × 0.128") or four 3" × 0.131" nails.

Double plate

Top plate

Stud

Nail Ceiling Joists, Rafters, and Ridge

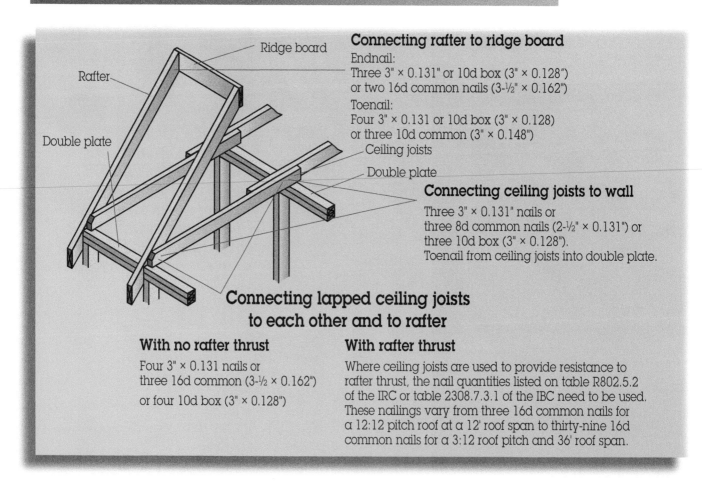

Ridge board

Rafter

Double plate

Connecting rafter to ridge board
Endnail:
Three 3" × 0.131" or 10d box (3" × 0.128")
or two 16d common nails (3-½" × 0.162")
Toenail:
Four 3" × 0.131 or 10d box (3" × 0.128)
or three 10d common (3" × 0.148")

Ceiling joists

Double plate

Connecting ceiling joists to wall

Three 3" × 0.131" nails or
three 8d common nails (2-½" × 0.131") or
three 10d box (3" × 0.128").
Toenail from ceiling joists into double plate.

**Connecting lapped ceiling joists
to each other and to rafter**

With no rafter thrust

Four 3" × 0.131 nails or
three 16d common (3-½ × 0.162")
or four 10d box (3" × 0.128")

With rafter thrust

Where ceiling joists are used to provide resistance to rafter thrust, the nail quantities listed on table R802.5.2 of the IRC or table 2308.7.3.1 of the IBC need to be used. These nailings vary from three 16d common nails for a 12:12 pitch roof at a 12' roof span to thirty-nine 16d common nails for a 3:12 roof pitch and 36' roof span.

Nail Rafters to Wall

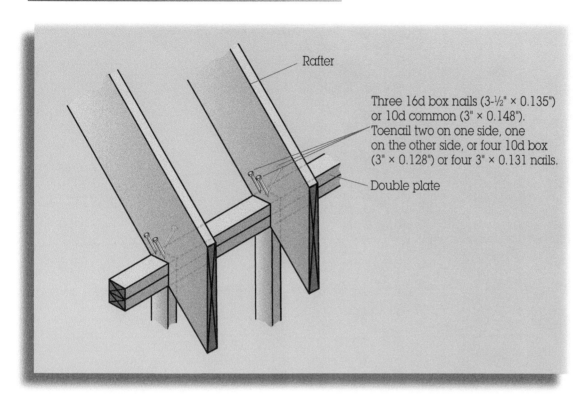

Rafter

Three 16d box nails (3-½" × 0.135")
or 10d common (3" × 0.148").
Toenail two on one side, one
on the other side, or four 10d box
(3" × 0.128") or four 3" × 0.131" nails.

Double plate

Nail Blocks

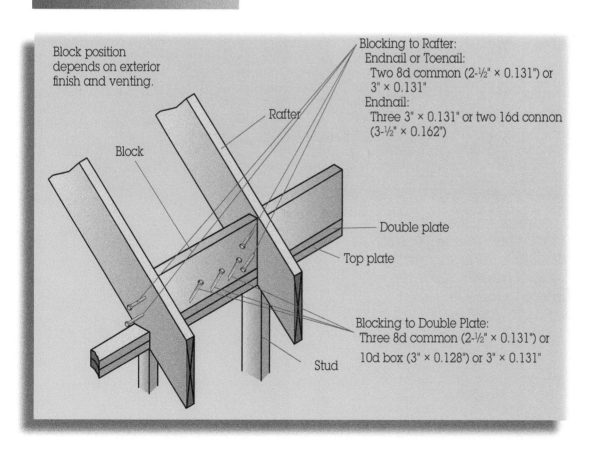

Block position
depends on exterior
finish and venting.

Rafter

Block

Blocking to Rafter:
  Endnail or Toenail:
    Two 8d common (2-½" × 0.131") or
    3" × 0.131"
  Endnail:
    Three 3" × 0.131" or two 16d connon
    (3-½" × 0.162")

Double plate

Top plate

Blocking to Double Plate:
  Three 8d common (2-½" × 0.131") or
  10d box (3" × 0.128") or 3" × 0.131"

Stud

Nail Fascia and Bargeboard

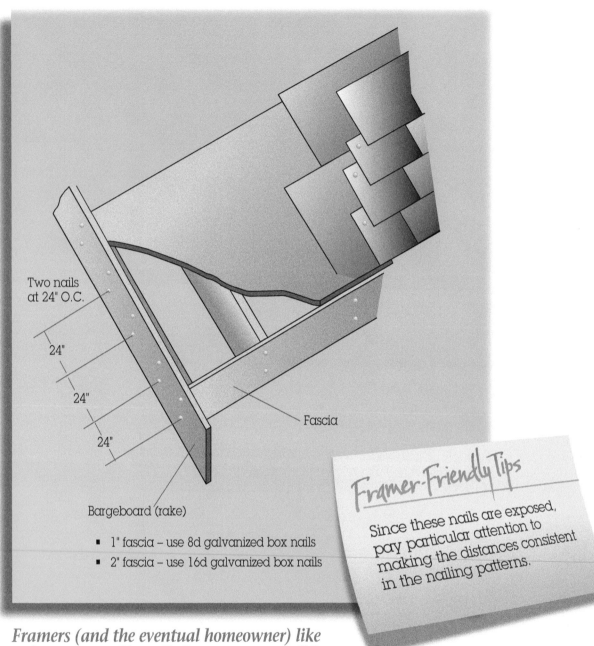

Two nails
at 24" O.C.

24"

24"

24"

Fascia

Bargeboard (rake)

- 1" fascia – use 8d galvanized box nails
- 2" fascia – use 16d galvanized box nails

Framer-Friendly Tips

Since these nails are exposed, pay particular attention to making the distances consistent in the nailing patterns.

Framers (and the eventual homeowner) like to have the roof installed as soon as possible. The roofing provides a dry workspace and protects all the installed framing from the weather. Be careful not to leave hammer head marks in the fascia, since it is a finish product.

Chapter Three
FLOOR FRAMING

Contents

Chapter Three

FLOOR FRAMING

This chapter illustrates the basic sequence for floor framing. Straight cuts and tight nailing make for a neat and professional job. Pay particular attention to the corners. It is important that they stay square and plumb up from the walls below, so the building does not gain or lose in size. Also pay close attention to laying the first sheet of subfloor sheathing. If it is laid straight and square, the entire subfloor will go down easily and you can avoid making extra cuts. If you make a sloppy start on the first sheet, you'll struggle to make each sheet fit, you'll waste valuable time, and you won't be proud of the results.

This chapter presents the 9 steps of floor framing.

Step 1-Crown and Place Joists

Spread joists so crowns are in the same direction. The crown is the highest point of a curved piece of lumber.

If the joists are resting on a foundation instead of a stud wall, then a sill plate, or mudsill, would be attached to the foundation, and the joists would rest on the plate or sill.

Framer-Friendly Tips

Marking the joists as you crown them–showing the crown direction–prevents mistakes or re-crowning.

Look for crown

Top

Mark joist crown up

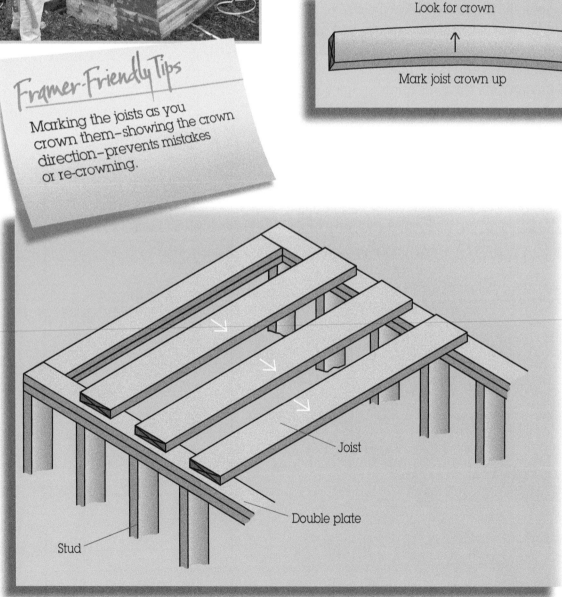

Joist

Double plate

Stud

Steps 2 and 3-Nail Rim Joists in Place and Cut Joists to Length

Joist spread and marked, ready to cut

If joists lap over an interior wall, they can be rough-cut approximately two inches beyond the wall. Do not let lapped joists go more than six inches beyond the wall.

Framer-Friendly Tips

When cutting lapped joists, check if the cut off can be cut the right length to use for blocking.

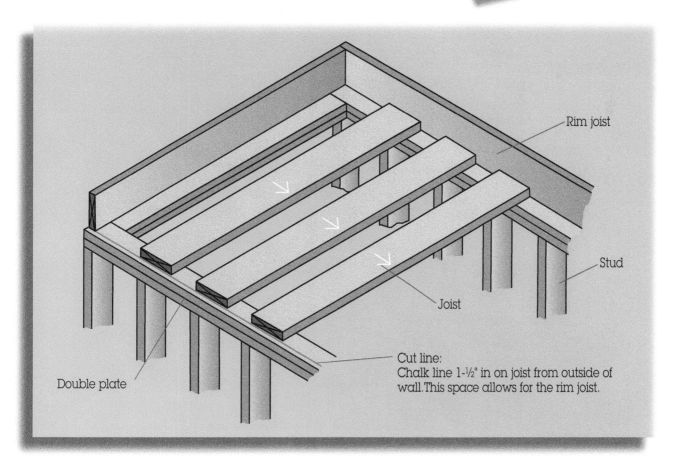

Rim joist

Stud

Joist

Cut line:
Chalk line 1-½" in on joist from outside of wall. This space allows for the rim joist.

Double plate

Step 4-Nail Joists in Place

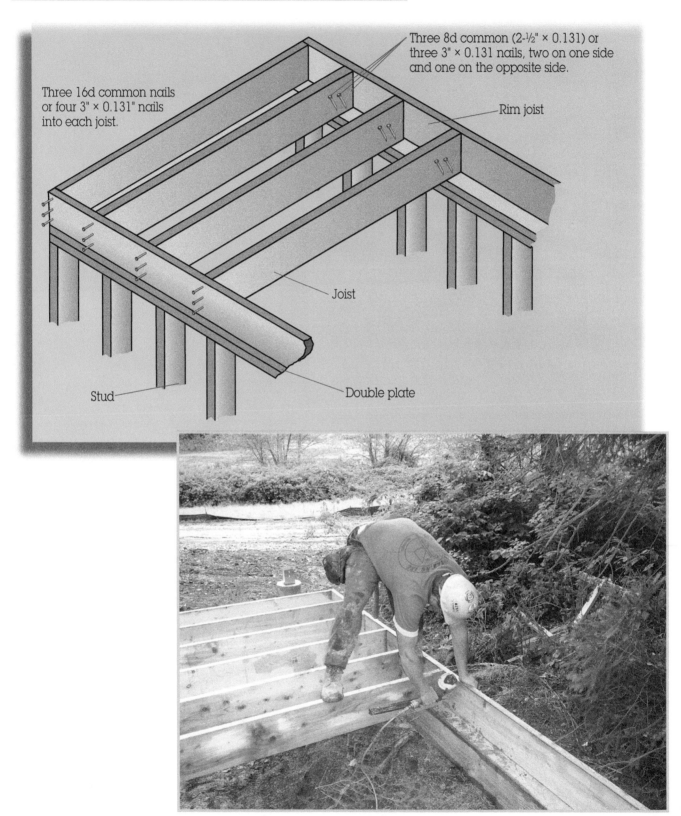

Three 8d common (2-½" × 0.131) or three 3" × 0.131 nails, two on one side and one on the opposite side.

Three 16d common nails or four 3" × 0.131" nails into each joist.

Rim joist

Joist

Stud

Double plate

Turn joists crown up and nail into place.

44

Step 5-Frame Openings in Joists

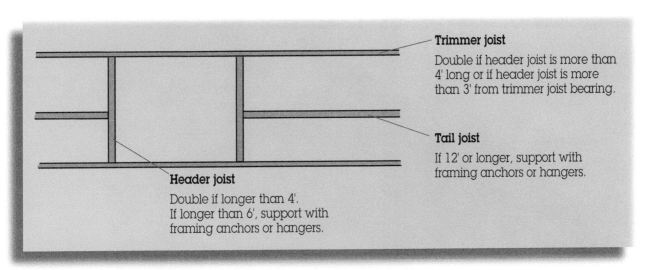

Trimmer joist

Double if header joist is more than 4' long or if header joist is more than 3' from trimmer joist bearing.

Tail joist

If 12' or longer, support with framing anchors or hangers.

Header joist

Double if longer than 4'. If longer than 6', support with framing anchors or hangers.

This is a crawl space access typically found in a closet or some other out-of-the-way location. Mark the floor sheathing as you install it so you don't cover the crawl space access.

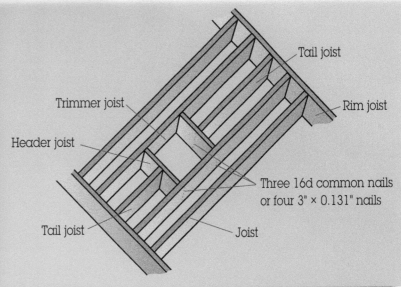

Tail joist

Rim joist

Trimmer joist

Header joist

Three 16d common nails or four 3" × 0.131" nails

Tail joist

Joist

Steps 6 and 7-Block Bearing Walls and Nail Joists to Walls

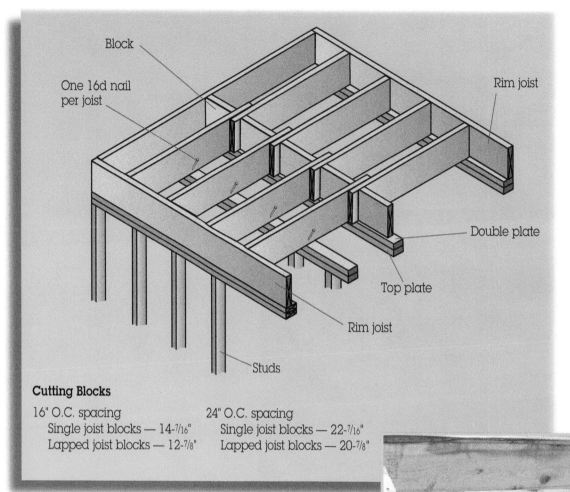

Block

One 16d nail per joist

Rim joist

Rim joist

Double plate

Top plate

Studs

Cutting Blocks

16" O.C. spacing
  Single joist blocks — 14-7/16"
  Lapped joist blocks — 12-7/8"

24" O.C. spacing
  Single joist blocks — 22-7/16"
  Lapped joist blocks — 20-7/8"

Framer-Friendly Tips

The right size block tightens up the joist location. Adjust block sizes as necessary.

Bearing wall blocks

Step 8-Drywall Backing

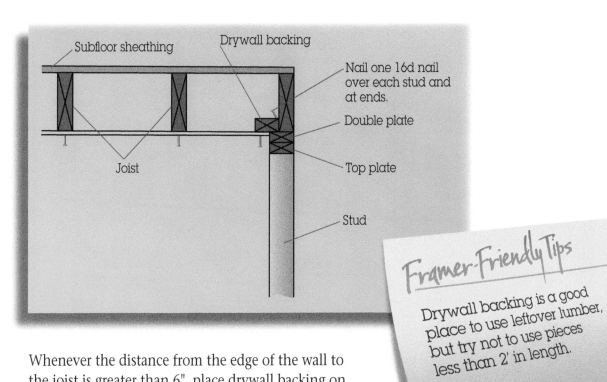

Whenever the distance from the edge of the wall to the joist is greater than 6", place drywall backing on top of the wall.

Framer-Friendly Tips

Drywall backing is a good place to use leftover lumber, but try not to use pieces less than 2' in length.

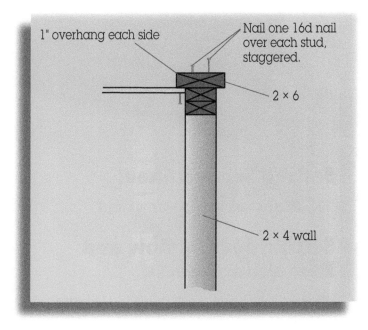

The width of the 2 × 6 provides for 1" of nailing surface on either side of the 2 × 4 wall.

Contents

Chapter Four

WALL FRAMING

There are many ways to frame walls, but it is always good to follow an organized sequence. This 16-step sequence has been developed over years of framing. Following these steps will help you and your crew work efficiently and eliminate errors. It will also ensure consistency from framer to framer. For example, if you have to leave a wall in the middle of framing it to go to another task, another framer can easily pick up where you left off and proceed without having to check every nail to see what you have done.

Keep in mind, walls must be square, plumb, and level. Measure accurately, cut straight, and nail tight.

Rake walls (sometimes referred to as *gable end walls*) typically start at the height of the standard wall and go up to the ridge of the roof. The challenge of building a rake wall is figuring the heights of the studs and making sure the wall is built square. Lifting the assembled rake wall into place can also be a challenge. This chapter will cover three ways to figure stud heights and build rake walls efficiently—using methods that will make your work easier.

Step 1-Spread Headers

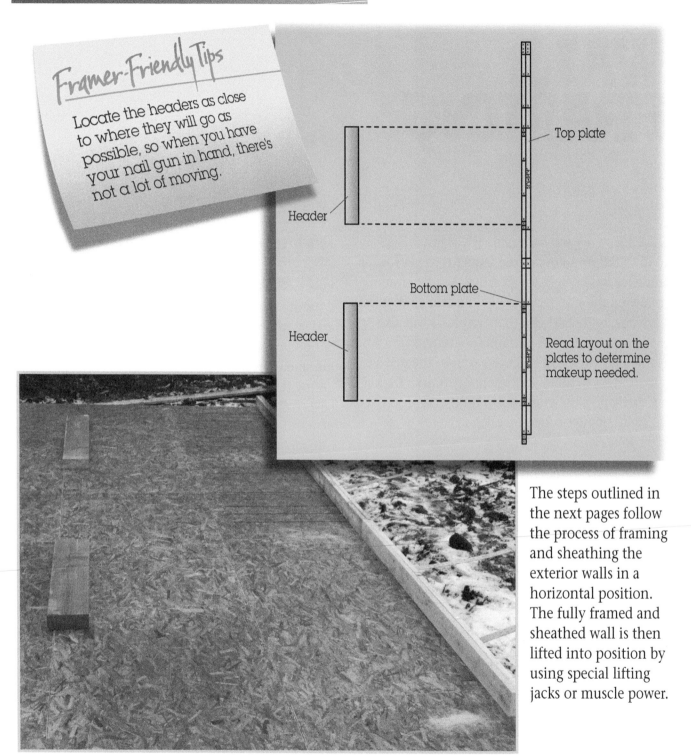

Framer-Friendly Tips

Locate the headers as close to where they will go as possible, so when you have your nail gun in hand, there's not a lot of moving.

Header

Header

Top plate

Bottom plate

Read layout on the plates to determine makeup needed.

The steps outlined in the next pages follow the process of framing and sheathing the exterior walls in a horizontal position. The fully framed and sheathed wall is then lifted into position by using special lifting jacks or muscle power.

Spread headers in location close to where they will be framed.

Step 2-Spread Makeup

Usually, one member of the framing crew does all the cutting. This promotes an efficient workflow and ensures consistent cuts. The person doing all the cutting must work fast enough to stay ahead of the remaining members of the crew.

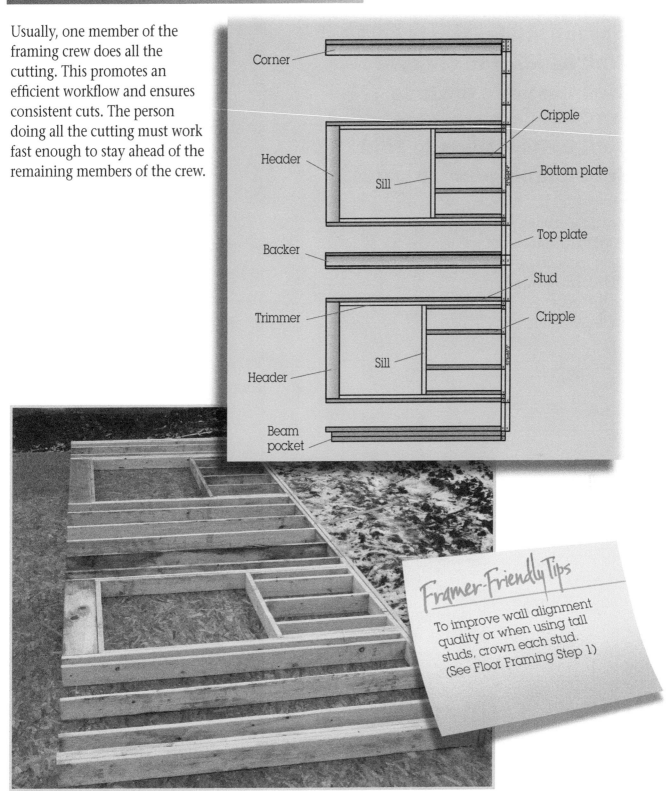

Spread makeup: stud/trimmers, backers, corners, cripples, sills, beam pockets.

Framer-Friendly Tips

To improve wall alignment quality or when using tall studs, crown each stud. (See Floor Framing Step 1)

Steps 3-7-Assemble Wall

3: **Spread studs**

4: **Nail headers to studs**

5: **Nail top plate to studs and headers**

6: **Nail bottom plate to studs**

7: **Nail double plate to studs**

Following the nailing patterns discussed in Chapter 2, the wall frame components can now be assembled. Pneumatic nailers reduce the amount of time required to perform this task.

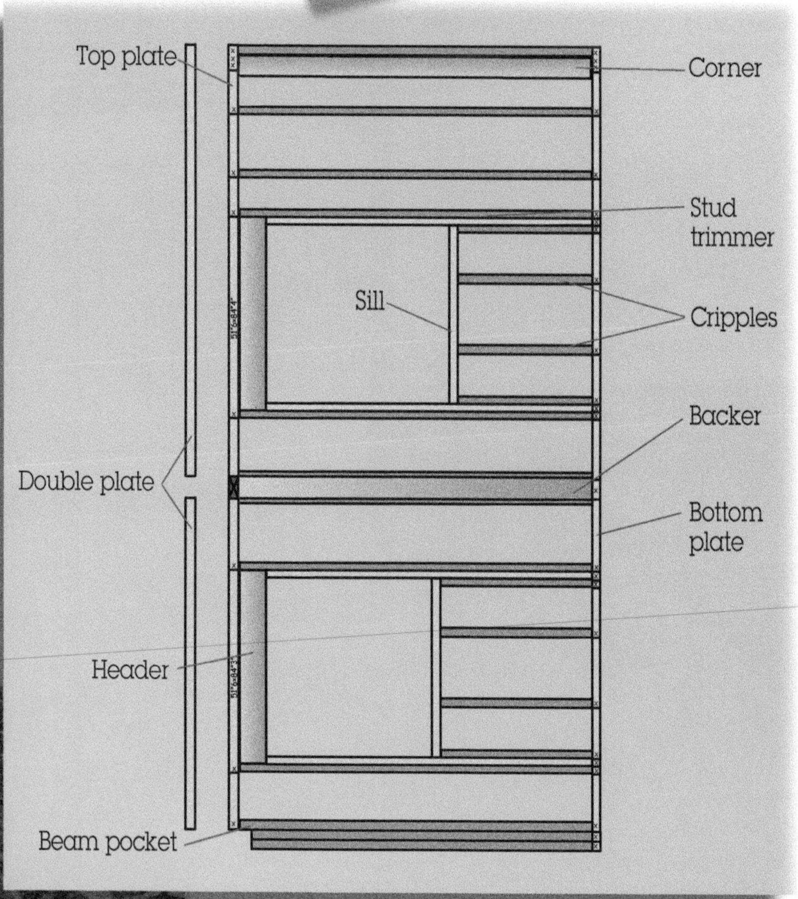

Framer-Friendly Tips

Check your window opening sizes once everything is nailed.

Top plate — Corner — Stud trimmer — Sill — Cripples — Backer — Double plate — Bottom plate — Header — Beam pocket

Spread remaining wall parts and nail.

Step 8–Square Wall

Exterior Wall

Framer-Friendly Tips

You can quickly figure how much you need to move the wall by finding the difference between the diagonals and taking half of that.

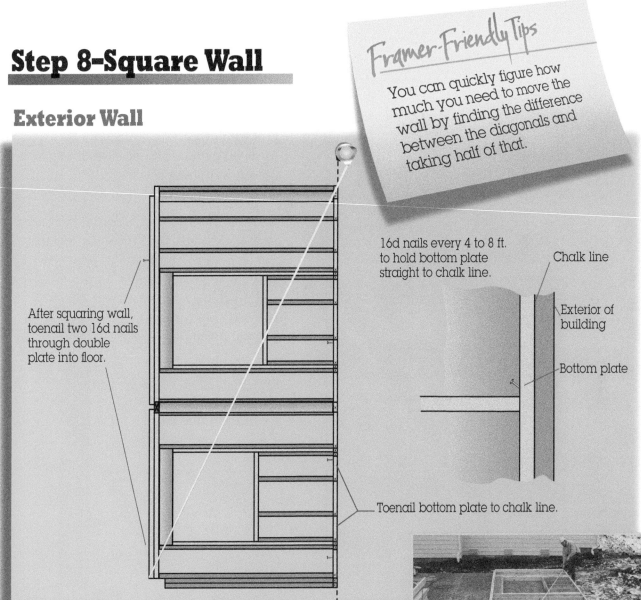

16d nails every 4 to 8 ft. to hold bottom plate straight to chalk line.

After squaring wall, toenail two 16d nails through double plate into floor.

Chalk line

Exterior of building

Bottom plate

Toenail bottom plate to chalk line.

To square a wall, secure the bottom plate as shown, then move the top of the wall until the diagonal dimensions are equal. Once the wall is square, secure it with two nails through the double plate into the floor.

Nail on the inside of the bottom plate so the nails will hold the wall in position while it is being stood. The bottom plate should be nailed so that it's in line with the wall chalk line. Then, when the wall is raised, it will be in the right position.

Step 9–Sheathe Wall

Sheathing: place, cut openings, and nail.

Cover the entire wall with sheathing, then rout window and door openings with a panel pilot router bit (see illustration). Save the leftover pieces of sheathing for small areas and filling in between floors.

If the first floor exterior walls can be reached from the ground, then the sheathing may not be installed until after the walls are plumb and lined (straight and true; see Step 16-Plumb and Line). This is one way to eliminate the potential problem of a square wall sitting on a foundation that is not level.

Panel pilot router bit

Wall sheathing

Openings for windows

Framer-Friendly Tips

If rafters or roof trusses will be resting on the corner of the wall, hold the sheathing down from the top – enough so that it will not interfere with the rafters.

Step 10-Install Nail-Flange Windows

Before Wall is Stood Up

a. Check plans for correct window.

b. Check window opening for protrusions (nails, wood splinters, etc.) that might hold window away from edge.

c. Install window flashing. (See "Window Flashing Installation," Chapter 6.)

d. Set window in opening, making sure window is right side up.

e. Slide window to each end of opening, and draw a line on the sheathing or flashing with a pencil along the edge of window. (Draw lines before caulking window.)

f. Caulk back side of header and jambs window flanges with ⅜" bead. Do not caulk sill. Caulk in line with prepunched nail holes.

g. Center window in marks you have just drawn.

h. Nail window sides and bottoms, using appropriate nails.

i. Do not nail top of window.

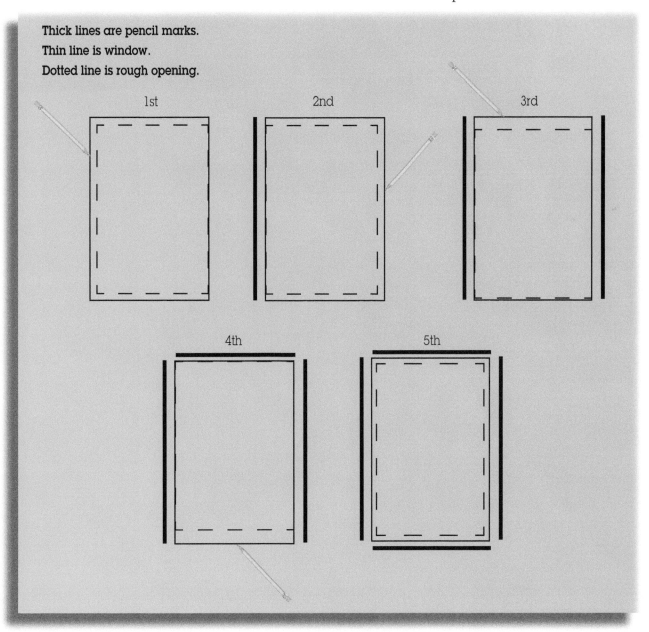

Thick lines are pencil marks.
Thin line is window.
Dotted line is rough opening.

1st   2nd   3rd

4th   5th

Steps 11-15-Standing and Setting Wall

11: **Stand wall.**

12: **Set bottom plate.**

13: **Set double plate.**

14: **Set reveal.**

15: **Nail wall.**

Framer-Friendly Tips

Before standing a wall, check underneath for miscellaneous debris, such as loose nails.

Corner securing two walls

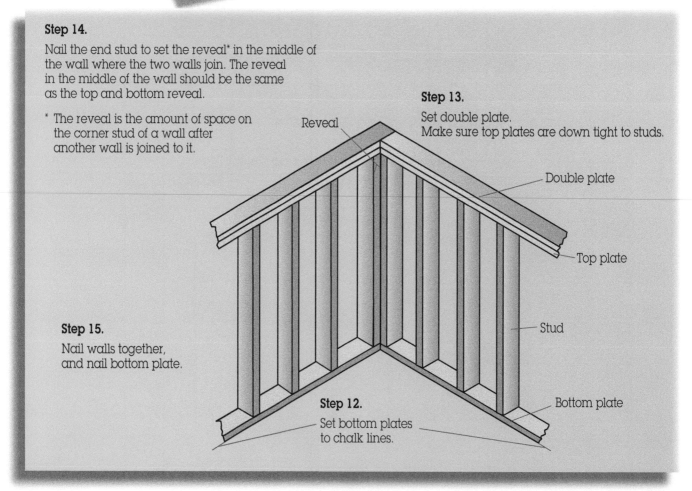

Step 14.

Nail the end stud to set the reveal* in the middle of the wall where the two walls join. The reveal in the middle of the wall should be the same as the top and bottom reveal.

* The reveal is the amount of space on the corner stud of a wall after another wall is joined to it.

Reveal

Step 13.

Set double plate.
Make sure top plates are down tight to studs.

Double plate

Top plate

Stud

Step 15.

Nail walls together, and nail bottom plate.

Bottom plate

Step 12.

Set bottom plates to chalk lines.

Step 16-Plumb and Line

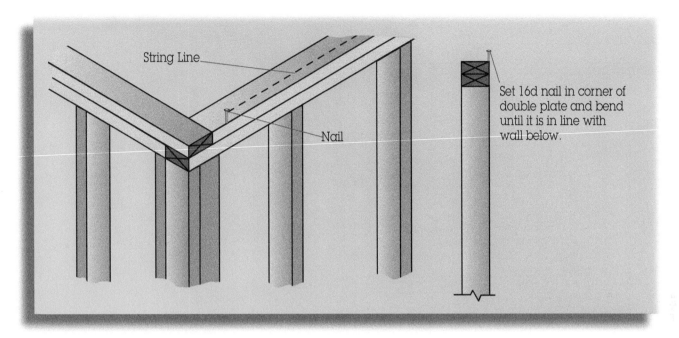

String Line

Nail

Set 16d nail in corner of double plate and bend until it is in line with wall below.

"Plumb and line" is the process of making the walls straight and true.

"Plumbing" is the use of a level to set the ends of the walls plumb or perfectly upright.

"Lining" is using a tight string attached to the top of a wall as a guide for straightening it.

Set nails at either end of wall as shown, and then string line tightly between them, adjusting the line so that it is about ½" above the double plate. Wall should be moved in or out to align with string.

The walls are braced with 2 × 4 lumber to hold them and, if necessary, make them plumb and straight.

If a wall already is sheathed and in place, but not plumb, correct it if it is more than ¼" out of plumb for standard height walls.

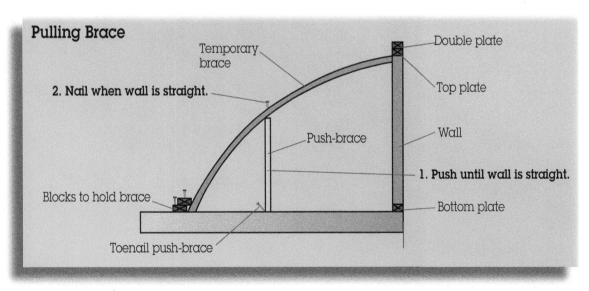

Pulling Brace

Temporary brace

2. Nail when wall is straight.

Double plate

Top plate

Push-brace

Wall

1. Push until wall is straight.

Blocks to hold brace

Bottom plate

Toenail push-brace

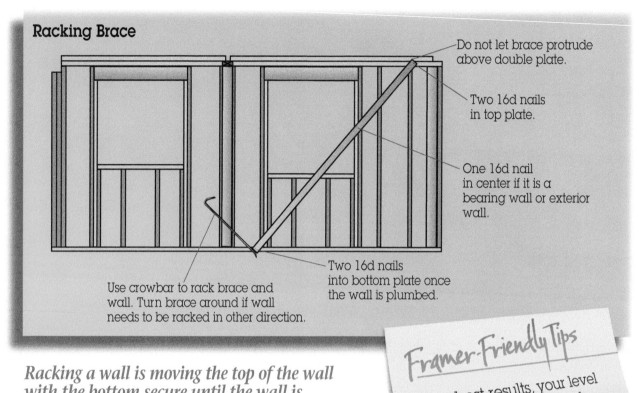

Racking Brace

Do not let brace protrude above double plate.

Two 16d nails in top plate.

One 16d nail in center if it is a bearing wall or exterior wall.

Two 16d nails into bottom plate once the wall is plumbed.

Use crowbar to rack brace and wall. Turn brace around if wall needs to be racked in other direction.

Racking a wall is moving the top of the wall with the bottom secure until the wall is plumb.

Framer-Friendly Tips

For best results, your level should be long enough to reach from bottom plate to top plate.

One framer racks the wall with a brace and crowbar, while the second framer checks for plumb.

Plumbing Tools

A level is the traditional tool for determining plumb. A laser can also be used. Whichever tool you use, you want to make sure you check it for true before you start. To check a level, hold it in position against a wall and read the bubble. Then turn it around and place it in the same position against the wall. If the bubble reads the same, then your level is accurate. To check a laser use a similar method. Set it in place, and mark the bottom dot and the top dot. Then turn it around and align the bottom dot. If the top dot hits the same spot on the top of the wall that you marked, your laser is reading accurately.

Levels are accurate when the level is long enough to reach from the bottom plate to the top plate. An 8' level will work fine for most residential walls. A 12' extension level works for walls up to 12'. If the level does not reach from plate to plate, you have to figure that there will be some variance for the studs' irregularities.

Lasers are good for taller walls. Lasers are not dependent on perfect wood for determining plumb. Set the laser at the bottom, and read the distance you set at the bottom on your tap at the top of the wall. If you need to check plumb on a regular basis, the laser has the advantage of fitting in your pouch.

Plumbing tall wall with extending level.

Framing Rake Walls

There are three common ways to figure stud heights and build rake walls.

1. Chalk lines on the floor
2. Figure lengths on paper
3. Chalk rake cut line

Each has its own advantages. With all these methods, you must use the pitch given in the plans to determine the wall height. The pitch is generally shown on the elevation sheet just above the roof slope.

Method 1: Chalk Lines on the Floor

The first method is to chalk out a duplicate on the floor, if you have the space. Then you can measure and cut the studs and plates right from your chalk lines.

The advantage to using this method is that it is quick, easy, accurate, and doesn't require a lot of math. However, if you don't have the space on the floor, if it's raining and you can't chalk lines, or you have a lot of rake walls in the building, it is probably best to use one of the other methods suggested.

To chalk the lines, you need to know the heights of your low point and high point. You must also ensure that the wall is square. To find the height, you can use the "Chalk Lines on the Floor" system (see illustration). The pitch on the plans gives you the relationship of the rise to the run. For example, a 6:12 pitch means that for every 12 units of run, there are 6 units of rise. To find the high point on the wall, go out 12 units of run, then up square 6 units. Mark this reference point and chalk a line from the low point

of the wall through this point, and extend it as far as necessary to reach the high point in the wall. The closer you make the reference point to the high point in the wall, the more accurate your line will be.

To find the rake wall stud heights using the "Chalk Lines on the Floor" method, follow these steps:

1. Chalk a bottom plate line (1). Usually you can use the chalk line for your wall. (See "Chalk Lines on the Floor" illustration.)
2. Chalk the short stud line (2). Make sure it's square (perpendicular) with the bottom plate line (1). You can use the 3-4-5 triangle to square the line. (Explained in "To Square the Wall" and "3-4-5 Triangle" illustration later in this chapter.)
3. Chalk a parallel line (3) with the bottom plate line (1) that aligns with the top of the short stud. Extend this line out toward the long stud (5).

An unsheathed rake wall

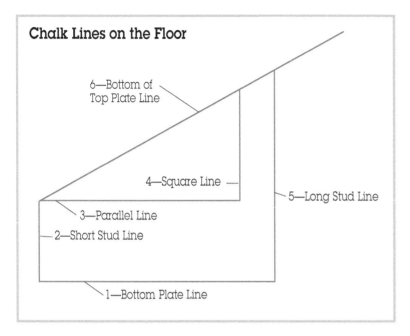

Chalk Lines on the Floor

6—Bottom of Top Plate Line

4—Square Line

5—Long Stud Line

3—Parallel Line

2—Short Stud Line

1—Bottom Plate Line

A 3-4-5 triangle will help you establish that two lines are square or at right angles to each other. To establish square, just follow these steps:

1. Start with the line you want to square from; this will be the 4-unit line—also referred to as the run.

2. Measure a line perpendicular to the run line at 3 units in length, called the rise.

3. Measure the diagonal from the outside of the 4-unit line (run) and the 3-unit line (rise), and adjust the 3-unit line so that the diagonal (hypotenuse) is exactly 5 units.

The units can be anything as long as they are in the same ratio. For example, they could be 3', 4', and 5', or they could be 15', 20', and 25'. The longer the units, the more accurate your measurement will be.

4. Chalk a square line (4) square to the parallel line (3). The length of the square line (4) will be in a relationship to the parallel line (3), depending on the pitch of the rake wall. If, for example, the pitch is 6:12 and the parallel line is 12, the square line will be 6.

5. Chalk a line (5) that is square with the bottom plate line (1) where the longest stud (5) should be.

6. Chalk a line from the short point of the short stud (2) through the top of the square line (4) and on past the long stud line (5). This will be your bottom of top plate line (6).

7. Once you have these lines, you will be able to fill in all the remaining studs.

To Square the Wall:

- Draw a straight line where you want to place your bottom plate, then make a perpendicular line at the high point of your wall.

- Use a 3-4-5 triangle to double-check that the line is exactly perpendicular or square. (See "3-4-5 Triangle" illustration.)

Method 2: Figure Lengths on Paper

With this method, you figure the stud heights, plate lengths, and layout anywhere you want—whether in the office, at home on a computer, or on the job site. All you need is a set of plans. Once you have the heights and lengths figured, you can build the wall anywhere, then move it into position.

If you are not using a computer, use the "Rake Wall Stud Heights" worksheet later in this chapter to figure the stud heights, the plate lengths, and the layout points. Give the completed worksheet with all needed information to whoever is framing the wall.

A construction calculator, such as Construction Master IV®, can be used to figure lengths accurately. With a construction calculator, you can work in feet and inches and use a memory function for repetitive calculations.

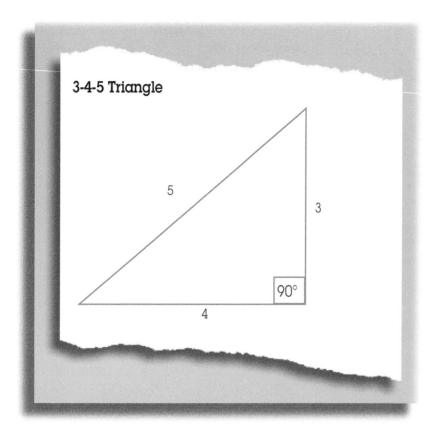

3-4-5 Triangle

5

3

4

90°

To use the Rake Wall Stud Heights Worksheet, just fill in the blanks and find the stud heights. The column "D" is where you write the distance on the bottom plate from the start of the rake wall to the short point on the stud. There is less confusion if you always use the short point on the studs. It is also easier to cut the short point than the long point when using a worm drive saw.

The stud height to the short point is found by using the formula (D × RP) + BH. RP is the rise percent, or the relationship between the rise and the run. The relationship gives you the height increase of the studs per increase in the distance of the plate. This relationship is illustrated in the filled-in version of the Rake Wall Stud Heights Worksheet. The formula for finding RP is also shown, in the "Rise and Diagonal Percent" illustration later in this chapter.

The "Rake Wall, RP, DP, Saw Angle" illustration provides the rise percent for common roof pitches. The "BH" from the formula is the beginning stud height. BH is a constant and is the height of the first stud at the lowest point. This height can vary depending on how the rafter or lookouts rest on the rake wall.

A typical beginning height would be slightly lower than the adjoining wall, as shown in the "Rake Wall Beginning Stud Height" drawing. In this example, the beginning stud height is only ⅜" less than the adjoining wall stud height because the plates on the rake are thicker on a slope than they are when flat.

To find the layout points for the studs and the length of the plates, use the formula D × DP. DP is the diagonal percent, or the relationship between the Diagonal and the Run. This relationship tells you the length increase of the top plate or layout point per increase in the distance of the bottom plate. This relationship is shown in the "Rise and Diagonal Percent" illustration, which also provides the formula for finding DP. The "Rake Wall, RP, DP, Saw Angle" illustration gives the Diagonal Percents for the common roof pitches and the saw angles—the different angles at which you can set your saw to cut the top of the studs and the ends of the top plate and double plate.

Framer-Friendly Tips

This method takes the most time to learn and set up, but once you master it, you'll probably use it exclusively.

Once you have completed the "Rake Wall Stud Heights" sheet, you'll have all the information you need to cut the studs and plates and lay out the plates. Then, just spread the studs and plates, and nail them together.

Rake Wall Stud Heights—Example

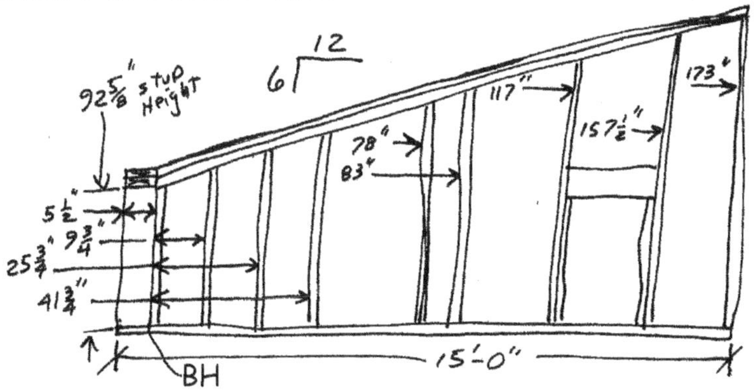

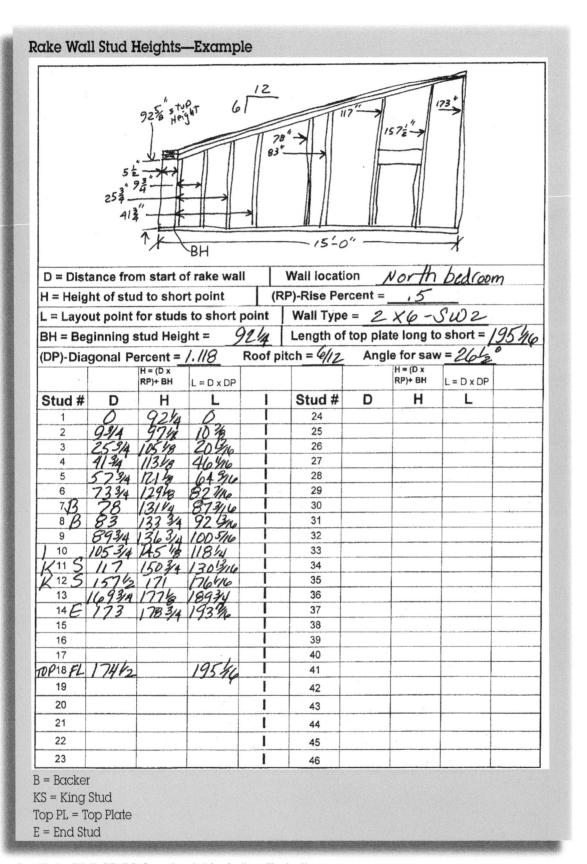

D = Distance from start of rake wall	Wall location North bedroom
H = Height of stud to short point	(RP)-Rise Percent = .5
L = Layout point for studs to short point	Wall Type = 2X6-SW2
BH = Beginning stud Height = 92 1/4	Length of top plate long to short = 195 1/16

(DP)-Diagonal Percent = 1.118 Roof pitch = 6/12 Angle for saw = 26 1/2°

		H = (D x RP)+ BH	L = D x DP				H = (D x RP)+ BH	L = D x DP
Stud #	**D**	**H**	**L**		**Stud #**	**D**	**H**	**L**
1	0	92 1/4	0		24			
2	9 3/4	97 1/8	10 7/8		25			
3	25 3/4	105 1/8	20 13/16		26			
4	41 3/4	113 1/8	46 11/16		27			
5	57 3/4	121 1/8	64 9/16		28			
6	73 3/4	129 1/8	82 7/16		29			
7 B	78	131 1/4	87 3/16		30			
8 B	83	133 3/4	92 13/16		31			
9	89 3/4	136 3/4	100 5/16		32			
10	105 3/4	145 1/8	118 1/4		33			
K 11 S	117	150 3/4	130 13/16		34			
K 12 S	157 1/2	171	176 1/16		35			
13	169 3/4	177 1/8	189 3/4		36			
14 E	173	178 3/4	193 7/16		37			
15					38			
16					39			
17					40			
TOP 18 PL	174 1/2		195 1/16		41			
19					42			
20					43			
21					44			
22					45			
23					46			

B = Backer
KS = King Stud
Top PL = Top Plate
E = End Stud

See "Rake Wall, RP, DP, Saw Angle" for further illustration

Method 3: Chalk Rake Cut Lines

This method is possibly the quickest way to figure stud heights and build rake walls. Here is how it is done:

- First, lay out the bottom plate in the same way you would if you were going to frame an ordinary wall.

- Spread your studs, making sure that they are long enough to reach the top of the rake wall.

- Toenail the bottom plate from the inside of the plate so that when the wall is lifted, the nail will function as a pivot point on the layout line.

- Cut the length of the beginning stud to match the adjoining wall. Take into consideration the location of the rafters if the lookouts rest on the rake wall, and the thickness of the plates on the rake. (See the "Rake Wall Beginning Stud Height" illustration.)

- Set the beginning stud square with the bottom plate.

- Use the rise percent to find the length of the longest stud. (See the "Rake Wall, RP, DP, Saw Angle" illustration.)

- Set that stud square with the bottom plate.

- Nail the rest of the studs to the bottom plate.

- Block the wall where required.

- Position all the studs so they are square.

- Chalk a line along the top of the studs.

- Cut each stud.

- Measure and cut the top plate and double plate.

- Nail the top plate to the studs, and the double plate to the top plate.

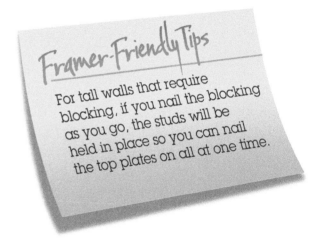

Framer-Friendly Tips

For tall walls that require blocking, if you nail the blocking as you go, the studs will be held in place so you can nail the top plates on all at one time.

Rake Wall Stud Heights Worksheet

D = Distance from start of rake wall	Wall location _____
H = Height of stud to short point	(RP)-Rise Percent = _____
L = Layout point for studs to short point	Wall Type = _____
BH = Beginning stud Height =	Length of top plate long to short =
(DP)-Diagonal Percent = _____ Roof pitch = _____ Angle for saw = _____	

Stud #	D	H = (D x RP)+ BH H	L = D x DP L	I	Stud #	D	H = (D x RP)+ BH H	L = D x DP L
1				I	24			
2				I	25			
3				I	26			
4				I	27			
5				I	28			
6				I	29			
7				I	30			
8				I	31			
9				I	32			
10				I	33			
11				I	34			
12				I	35			
13				I	36			
14				I	37			
15				I	38			
16				I	39			
17				I	40			
18				I	41			
19				I	42			
20				I	43			
21				I	44			
22				I	45			
23				I	46			

B = Backer
KS = King Stud
Top PL = Top Plate

Framer-Friendly Tips
You can copy this page for your own walls, or use a spreadsheet on your digital pad to create your own.

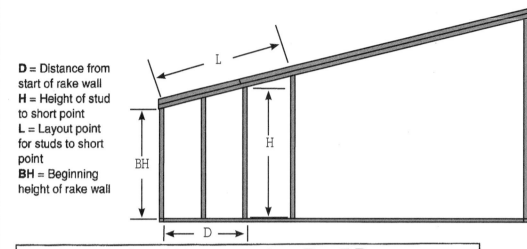

D = Distance from start of rake wall
H = Height of stud to short point
L = Layout point for studs to short point
BH = Beginning height of rake wall

$$H = (D \times RP) + BH \qquad L = D \times DP$$

PITCH	RP-Rise Percent	DP-Diagonal Percent	SAW ANGLE			
1/12	0.08	1.00	4.50			
2/12	0.17	1.01	9.50			
3/12	0.25	1.03	14.00			
4/12	0.33	1.05	18.50			
5/12	0.42	1.08	22.50			
6/12	0.50	1.12	26.50			
7/12	0.58	1.16	30.25			
8/12	0.67	1.20	33.75			
9/12	0.75	1.25	37.00			
10/12	0.83	1.30	40.00			
11/12	0.92	1.36	42.50			
12/12	1.00	1.41	45.00			

Rake Wall Beginning Stud Height

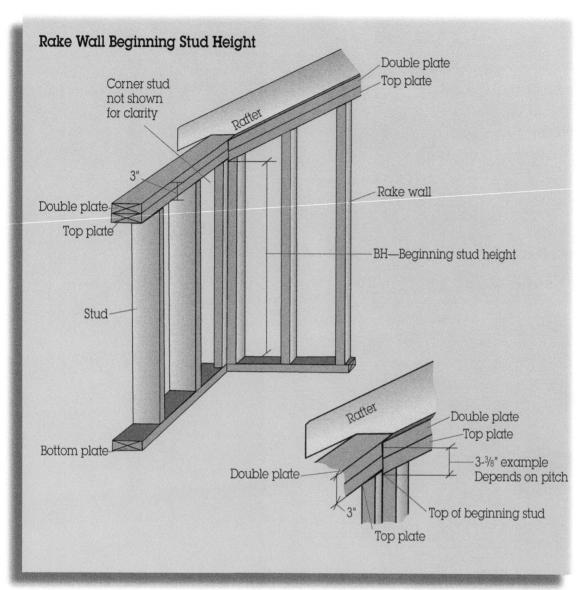

Corner stud not shown for clarity

Double plate
Top plate
Rafter
3"
Double plate
Top plate
Rake wall
BH—Beginning stud height
Stud
Bottom plate

Rafter
Double plate
Top plate
Double plate
3-³⁄₈" example
Depends on pitch
3"
Top of beginning stud
Top plate

Please note that the beginning stud height on a rake wall is not the same as a typical stud.

Rise and Diagonal Percent

Rise Percent

RP = Rise Percent = Rise divided by Run = $^{Rise}/_{Run}$

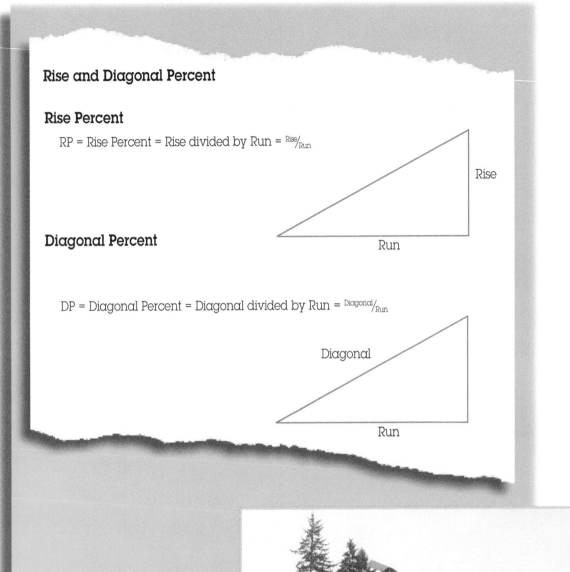

Diagonal Percent

DP = Diagonal Percent = Diagonal divided by Run = $^{Diagonal}/_{Run}$

Standing a rake wall

Chapter Five
ROOF FRAMING

Contents

Chapter Five

ROOF FRAMING

Framing a roof is the most difficult aspect of framing. The ability to construct a roof is a real test of your framing skill.

This chapter starts with a review of important rafter terms and roof styles, then it presents the eleven steps for framing a roof with rafters. After the eleven steps are presented, an example walks you through cutting rafters on a complex roof. The last part of this chapter discusses ceiling joists and then presents eight steps of framing a roof using roof trusses.

The example of cutting rafters covers finding rafter lengths using the "Diagonal Percent" method. You can follow this example to learn how to cut difficult rafters. If you master this system, you will be able to "Cut and Stack" a roof, which means you can cut all the rafters and stack them on the ground ready for installation before the first one is installed.

Roof trusses are typically pre-manufactured off site and delivered to the job site. They are engineered for strength and use standard dimension lumber. Gang nail plates are used to connect the cords and struts that make up the trusses. Uniform in size, they are somewhat easier to work with than ridge boards and rafters. Still they are heavy to work with, especially working at heights.

The most difficult part of framing a roof is finding the rafter length. It is based on the relationship between the rise (vertical) distance and the run (horizontal) distance. In this chapter, you will see different methods for finding the rafter length, though you will only need to memorize the approach you are most comfortable with. You will also find that at different times you might want to use different methods.

The angles and pitches of a roof are as varied as the colors in a child's crayon box. Just as some colors have certain characteristics in common, so do rafters and trusses. This chapter is organized around these common characteristics and these characteristics are presented in the eleven steps for framing a roof with rafters and the eight steps for framing a roof with trusses.

Roof Framing Terms

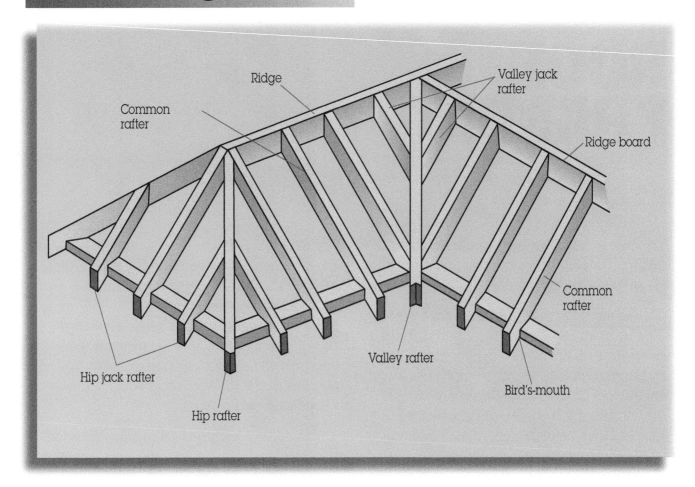

Ridge

Common rafter

Valley jack rafter

Ridge board

Common rafter

Hip jack rafter

Valley rafter

Bird's-mouth

Hip rafter

Important Rafter Terms

Span—the distance between two supporting members, typically measured from the outside of two bearing walls.

Run—horizontal distance.

Rise—vertical distance.

Diagonal—the distance between the far point on the run and the high point on the rise. (Similar to hypotenuse in mathematical terms.)

Hip or valley run—the horizontal distance below the hip or valley of a roof, from the outside corner of the wall to the center framing point.

Overhang hip run—the horizontal distance below the overhang hip of the roof, from the outside corner of the wall to the outside corner of the fascia.

Hip or valley diagonal—the distance between the far point on the hip or valley run and the high point on the hip or valley rise.

Overhang diagonal—the distance between the far point on the overhang run and the high point on the overhang rise.

Diagonal percent—the diagonal divided by the run.

Hip-val diagonal percent—the hip or valley diagonal divided by the hip or valley run.

Rise percent—the rise divided by the run.

Pitch—the slope of the roof, or the relationship of the run to the rise. Typically defined as a certain height of rise for 12 units of run for a common rafter, and 17 (16.97) units of run on a 90° hip or valley rafter.

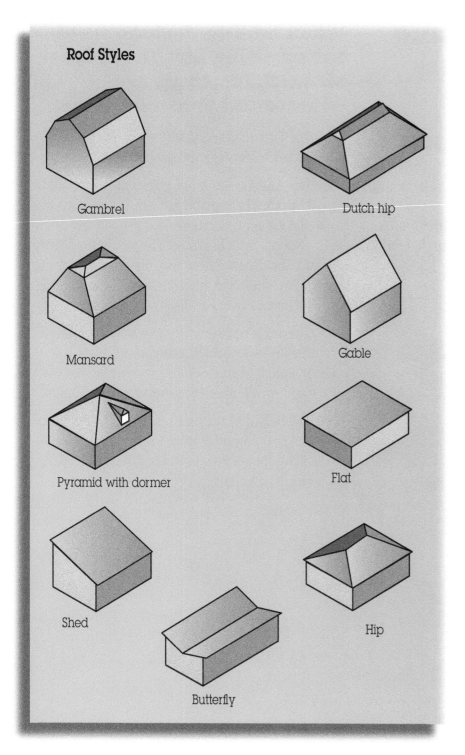

Roof Styles

Gambrel

Dutch hip

Mansard

Gable

Pyramid with dormer

Flat

Shed

Butterfly

Hip

Framing point—the point where the center lines of connecting rafters, ridges, hips, or valleys would meet.

Cheek cut—an angle cut that is made to bear against another rafter, hip, or valley.

Common rafter—a rafter running from a wall straight to a ridge board.

Jack rafter—a rafter running to a hip rafter or a valley rafter.

Hip rafter—a rafter at an outside corner of a roof that runs in between and joins jack rafters that bear on corner walls.

Valley rafter—a rafter at an inside corner of a roof that runs between and joins with jack rafters from each side.

Ridge end rafter—a rafter that runs from the end of a ridge.

Pitch angle—the vertical angle on the end of a rafter that represents the pitch of the roof.

Connection angle—the horizontal angle at the end of a rafter needed to connect to other rafters, hips, valleys, or ridge boards.

75

11 Steps for Framing a Roof with Rafters

Step 1-Find the Lengths of Common Rafters

There are six methods for finding common rafter lengths. Study them all and use the one that works best for you.

They are:

A. The Pythagorean Theorem
B. Framing-Square Rafter Method
C. Framing-Square Stepping Method
D. Chalking Lines Duplication Method
E. Computer Software Method
F. Diagonal Percent Method

When finding the lengths of rafters using any one of these six methods, there are 4 important considerations:

1. The length: determined by two factors—distance spanned and slope.

2. The adjustment to the length at the top and bottom. The top and bottom adjustments can depend on a number of factors and are almost always a little different. The two main factors are the distance from the true ridge or framing point, and the connection with other framing members.

3. The angle of cuts at the top, bottom, and the bird's mouth. The angle cuts relate to the pitch of the roof and the position of the framing member the rafters are attaching to.

4. The height at the bird's mouth. The height of the bird's mouth can be set by details on the plans, for bearing, or to keep the roof level at the plate height.

Methods for Finding Rafter Lengths

A. Pythagorean Theorem

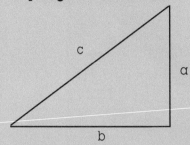

B. Framing-Square Rafter Method

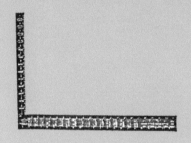

C. Framing-Square Stepping Method

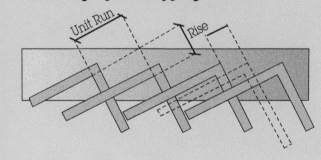

D. Chalking Lines Duplication Method

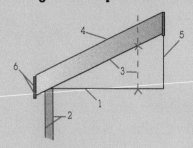

E. Computer Software Method

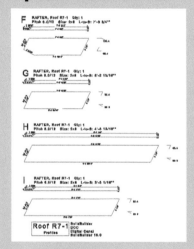

F. Diagonal Percent Method

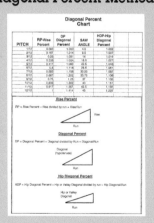

77

A. Pythagorean Theorem

Pythagoras was an ancient Greek philosopher and mathematician. His famous theorem states that the square of the hypotenuse of a right triangle is equal to the sum of the squares of the two other sides.

Thus: $A^2 + B^2 = C^2$

In roof framing:

A = the Rise

B = the Run

C (the hypotenuse) = the Rafter Length.

Run = ½ building width – ½ ridge board width

H = is given on plans = the amount of rise per foot of run

Rafter Cut Length = Rafter Length + Rafter Tail Length

Finding Rafter Length

First, find the run by using this formula:

Run = ½ building width – ½ ridge board width

Second, find the rise by using this formula:

Rise = $^H/_{12}$ × Run

Third, find the rafter length by using this formula:

Rafter Length = $\sqrt{(\text{Rise} \times \text{Rise}) + (\text{Run} \times \text{Run})}$

Framer-Friendly Tips

If math is not your strong point, skip this method and try the "chalking lines duplication" method.

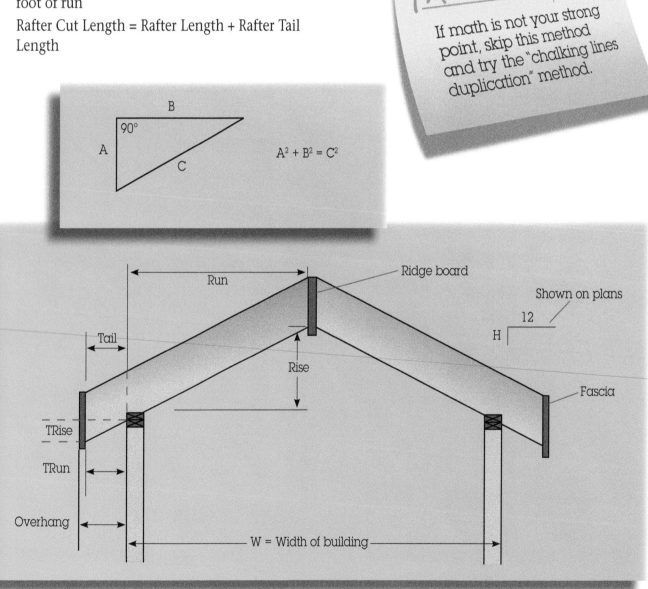

A. Pythagorean Theorem (continued)

To apply this formula, multiply Rise × Rise, and then Run × Run. Add the two products, then press the square root key on your calculator. The result is the Rafter Length.

Finding Rafter Tail (T) Length

First, find the TRun by using the following formula:

TRun = Overhang – Fascia

Second, find the TRise using the following formula:

TRise = $^H/_{12}$ × TRun

Third, find the Rafter Tail Length by using the following formula:

Rafter Tail Length = $\sqrt{(\text{TRise} \times \text{TRise}) + (\text{TRun} \times \text{TRun})}$

Note: Be sure to mark crowns on rafters prior to measuring and cutting. Crowns are always up.

Example: Finding Rafter Cut Length

Rafter Length: Let the pitch be $4\frac{12}{}$ and the building width be 20' and the ridge board be 1-½" thick.

Step 1: Run	= ½ (20') – ½ (1-½") = 9'-11-¼"
Step 2: Rise	= $^4/_{12}$ × 9'-11-¼"
	= 0.3333 × 119.25"
	= 39.75
	= 39-¾"
Step 3: Rafter Length	= $\sqrt{(119.25 \times 119.25) + (39.75 \times 39.75)}$
	= $\sqrt{14{,}220.56 + 1{,}580.06}$
	= $\sqrt{15{,}800.62}$
	= 125.70
	= 125-$^{11}/_{16}$"

Rafter Tail Length: Let overhang be 2' and fascia be 1-½".

Step 1: TRun	= 2'-1-½" = 1'-10-½"
Step 2: TRise	= $^4/_{12}$ × 1'-10-½"
	= 0.3333 × 22.5" = 7.49
	= 7.49
	= 7-½"
Step 3: Rafter Tail Length	= $\sqrt{(7.49 \times 7.49) + (22.5 \times 22.5)}$
	= $\sqrt{56.10 + 506.25}$
	= $\sqrt{562.35}$
	= 23.71
	= 23-$^{11}/_{16}$"
Rafter Cut Length	= 125-$^{11}/_{16}$" + 23-$^{11}/_{16}$"
	= 149-$^3/_8$"

TRun = length of a horizontal line from the building's exterior wall to the inside of the fascia. (See diagram on page 78.)

TRise = Amount of vertical rise in the length of TRun.

D. Chalking Lines Duplication Method

This method is probably the easiest to use. To find the lengths of rafters, you make an actual size drawing of the rafter on the floor and then measure the length.

All the information you need to use this method should be on the plans. First you will need the *pitch*. (See Method C mentioned previously.) Second is the *span*, which is the distance from the support on one side of the rafter to the support on the other side. Third is the *width* of the rafter, the length of the roof overhang, and the size of the exterior wall.

Steps (shown in Line Chalking Sequence illustration):

1. Chalk a straight line longer than the length of your rafter, which would represent the bottom of the ceiling joist if there were a ceiling joist.

2. Chalk two lines perpendicular to the first line to represent the exterior wall.

3. From the point where your inside exterior wall line crosses the ceiling joist line, measure out and up according to your pitch $\times \frac{12}{\Gamma}$. For this example let your pitch equal $6\frac{12}{\Gamma}$. Therefore, for every 12" you measure out, you measure 6" up. The longer the distance out, the greater your accuracy. Make sure that the line up is exactly perpendicular or square. You can use surrounding walls that are square to measure from or use a 3-4-5 triangle.

4. Chalk a line for the thickness of the rafter.

5. Measure the distance for the span along the ceiling joists line, then make a perpendicular line up and mark the ridge board.

6. Measure the distance of the roof overhang and draw in the fascia board.

With these lines in place, you can measure all the lengths you will need to cut your rafters.

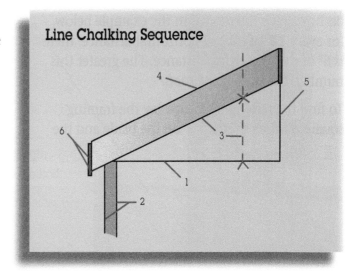

Chalking rafter lines

Roof Production

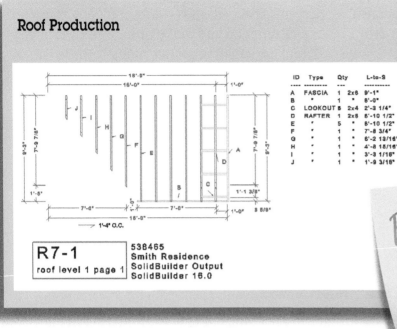

ID	Type	Qty		L-to-S	bevel	miter
A	FASCIA	1	2x6	9'-1"	--See profile--	
B	"	1	"	6'-0"	--See profile--	
C	LOOKOUT	5	2x4	2'-3 1/4"		
D	RAFTER	1	2x6	6'-10 1/2"	--See profile--	
E	"	5	"	6'-10 1/2"	--See profile--	
F	"	1	"	7'-8 3/4"	--See profile--	
G	"	1	"	6'-2 13/16"	--See profile--	
H	"	1	"	4'-8 15/16"	--See profile--	
I	"	1	"	3'-3 1/16"	--See profile--	
J	"	1	"	1'-9 3/16"	--See profile--	

R7-1
roof level 1 page 1

538465
Smith Residence
SolidBuilder Output
SolidBuilder 16.0

Framer-Friendly Tips

Entering the information into a computer is time-consuming, but some software designed for architects can save time by producing rafter cut calculations.

Rafter Profiles

F RAFTER, Roof R7-1 Qty: 1
Pitch 6.0/12 Size: 2x6 L-to-S: 7'-8 3/4"

G RAFTER, Roof R7-1 Qty: 1
Pitch 6.0/12 Size: 2x6 L-to-S: 6'-2 13/16"

H RAFTER, Roof R7-1 Qty: 1
Pitch 6.0/12 Size: 2x6 L-to-S: 4'-8 15/16"

I RAFTER, Roof R7-1 Qty: 1
Pitch 6.0/12 Size: 2x6 L-to-S: 3'-3 1/16"

Roof R7-1
Profiles

SolidBuilder
DCC
Digital Canal
SolidBuilder 16.0

E. Computer Software Method

Using the methods described on the previous page to find the lengths and angles for cutting rafters is not easy, but it is at least organized—and with a calculator that works in feet and inches and that figures the diagonals automatically, the process is straightforward. However, the easiest method is to use the computer. There are software programs currently available that will do all the work for you and produce a sketch of each rafter. Solid Builder is one of these programs. The illustrations on this page were done in Solid Builder. "Roof Production" identifies the type of roof parts, the quantity, lumber, and strength. "Rafter Profiles" illustrates the individual rafters with the balance of information you will need for cutting the rafters.

The hardest part of producing these computer-generated diagrams is learning the software and then entering the information needed for each structure in order to generate the diagrams. However, for the architect who has already drawn up the plans, or the builder who is working with computer-generated plans, it is an easy task to produce these rafter profiles. If computer-generated rafter profiles were prepared and attached to plans, it could really make framing roofs a breeze.

F. Diagonal Percent Method

The length of a rafter can be found by determining the horizontal length (run) that it covers, and the pitch of the roof. The constant relationship between these factors is defined as the **diagonal percent**. This percent is constant for any common or jack rafter on any roof that has the same pitch. To find the length of a rafter, multiply the length of the run by the diagonal percent. For example, if you have a roof with a $^6/_{12}$ pitch and a run of 6'-11-¼", you multiply 6'-11-¼" by 1.118 (diagonal percent for a $^6/_{12}$ pitch) and find that your rafter length is 7'-9-¹⁄₁₆". With a construction calculator, enter 6'-11-¼" × 1.118, and it will read 7'-9-¹⁄₁₆". The illustration below provides the diagonal percent for common pitch roofs. To figure the length of hip and valley rafters, use the hip-val diagonal percent shown on the chart.

When you use the diagonal percent, the most difficult part of figuring rafter length is finding the length of the run. The adjustments that need to be made at the top and bottom of the rafter should be added and subtracted from the run before the rafter length is calculated. In finding the run, it is best to start with the full run distance from the outside of the bearing wall to the framing point of any connecting framing member. Use the framing point for consistency, and then make adjustments from there.

Diagonal Percent Chart

PITCH	RP-Rise Percent	DP Diagonal Percent	SAW ANGLE	HDP-Hip Diagonal Percent		
1/12	0.083	1.003	4.5	1.002		
2/12	0.167	1.014	9.5	1.007		
3/12	0.25	1.031	14	1.016		
4/12	0.333	1.054	18.5	1.027		
5/12	0.417	1.083	22.5	1.043		
6/12	0.5	1.118	26.5	1.061		
7/12	0.583	1.158	30.25	1.082		
8/12	0.667	1.202	33.75	1.106		
9/12	0.75	1.25	37	1.132		
10/12	0.833	1.302	40	1.161		
11/12	0.917	1.357	42.5	1.192		
12/12	1	1.414	45	1.225		

Rise Percent

RP = Rise Percent = Rise divided by run = Rise/Run

Diagonal Percent

DP = Diagonal Percent = Diagonal divided by Run = Diagonal/Run

Hip Diagonal Percent

HDP = Hip Diagonal Percent = Hip or Valley Diagonal divided by run = Hip Diagonal/Run

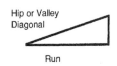

Framer-Friendly Tips

To get the rafter's length, measure the run and multiply it by the roof's diagonal pitch percent.

Step 2-Cut Common Rafter

The illustrations in Step 2 show details on marking:

   A. Rafter cut length

   B. Bird's-mouth cuts

   C. Angle cuts

A. Steps for marking rafter length

1. Mark top and bottom of rafter (see step #1, "Find the Lengths of Common Rafters").

2. Set speed square so pitch mark aligns with edge of rafter.

3. Mark line along edge of speed square at top and bottom.

Cut a pattern first and try it for fit before cutting all the rafters. A framing square can also be used to mark cut lines. (See illustration in "Framing-Square Stepping Method" earlier in this chapter.)

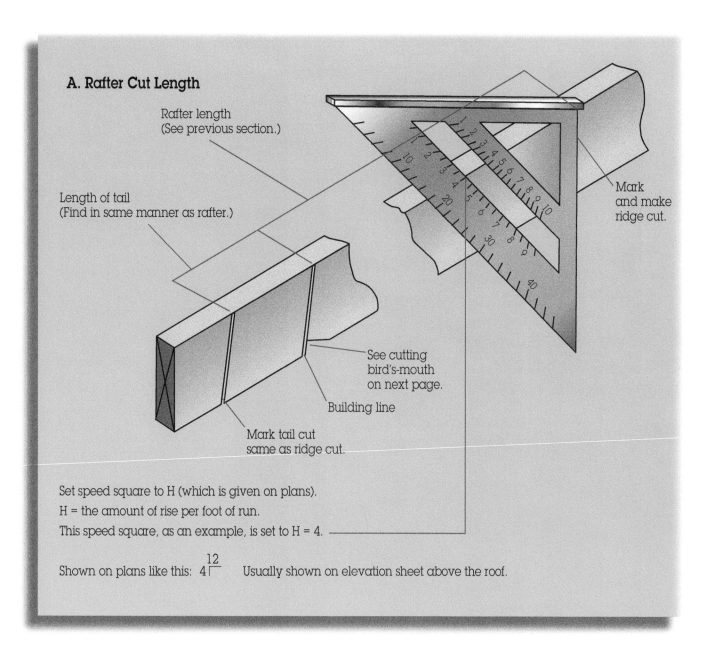

A. Rafter Cut Length

Rafter length
(See previous section.)

Length of tail
(Find in same manner as rafter.)

Mark and make ridge cut.

See cutting bird's-mouth on next page.

Building line

Mark tail cut same as ridge cut.

Set speed square to H (which is given on plans).

H = the amount of rise per foot of run.

This speed square, as an example, is set to H = 4.

Shown on plans like this: 4 $\frac{12}{}$ Usually shown on elevation sheet above the roof.

Step 2-Cut Common Rafter (continued)

B. Steps for marking bird's-mouth cuts

1. Mark rafter length.

2. Mark building line at rafter length for the correct pitch.

3. Mark parallel plumb line a distance equal to width of wall and toward interior of building.

4. Mark seat cut square (90°) from building line at rafter length and bottom of parallel line.

5. Cut bird's-mouth.

A framing square can also be used to mark a bird's-mouth. (See illustration in "Framing-Square Stepping Method" earlier in this chapter.)

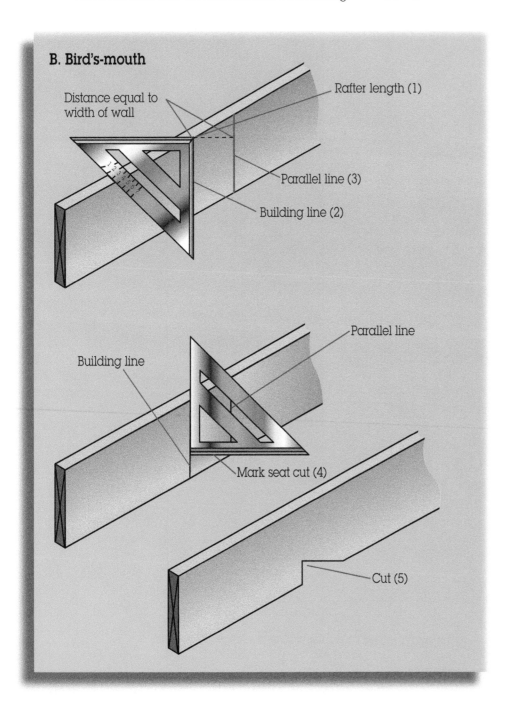

B. Bird's-mouth

Distance equal to width of wall

Rafter length (1)

Parallel line (3)

Building line (2)

Parallel line

Building line

Mark seat cut (4)

Cut (5)

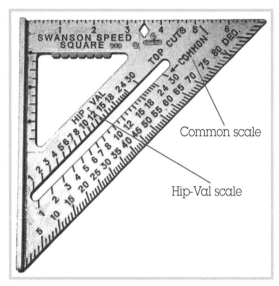

Use numbers on speed square to match pitch on roof

C. Angle Cuts Basics

1. Use the common scale on the speed square to mark and cut common rafters and hip jack and valley jack rafters.
2. Use the angle set on your saw to cut 90° square for top and bottom of common rafters and top of valley jack rafters and bottom of hip jack rafters.
3. When hip and valley corners are at 90° set your saw at 45° to cut the top of hip jack rafters and the bottom of valley jack rafters.
4. Use the Hip-Val scale on the speed square to make a line and cut the top and bottom of the hip or valley rafters.
5. When hip and valley corners are at 90° set the saw at 45° and cut ½ the rafter.
6. The 45° saw angle will change if the angles of the corners change.

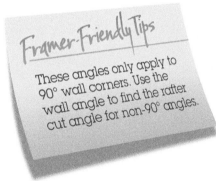

Framer-Friendly Tips

These angles only apply to 90° wall corners. Use the wall angle to find the rafter cut angle for non-90° angles.

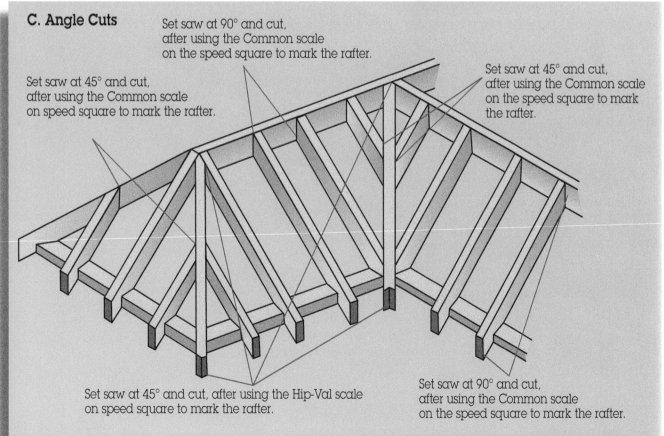

C. Angle Cuts

Set saw at 90° and cut, after using the Common scale on the speed square to mark the rafter.

Set saw at 45° and cut, after using the Common scale on speed square to mark the rafter.

Set saw at 45° and cut, after using the Common scale on the speed square to mark the rafter.

Set saw at 45° and cut, after using the Hip-Val scale on speed square to mark the rafter.

Set saw at 90° and cut, after using the Common scale on the speed square to mark the rafter.

Step 3–Set Ridge Board

1. Lay temporary sheathing under where ridge board will be.

2. Figure length and set temporary ridge support.
3. Install ridge board.

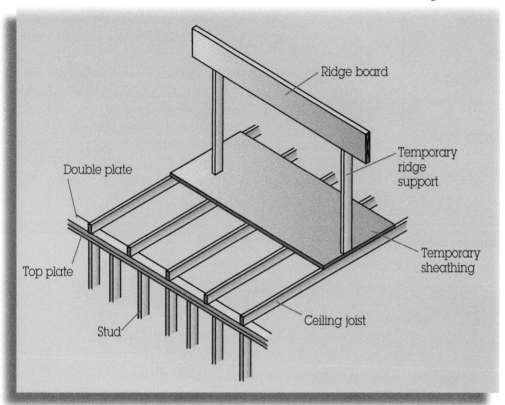

Ridge board ready for rafters

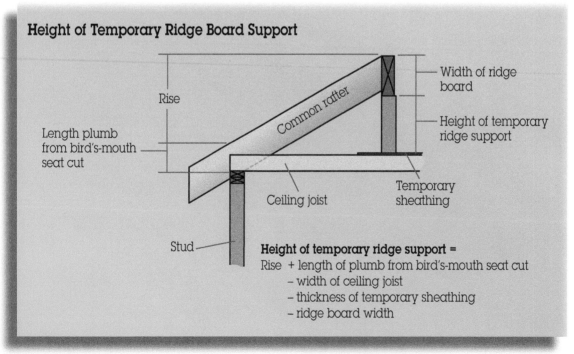

Height of Temporary Ridge Board Support

Height of temporary ridge support =
Rise + length of plumb from bird's-mouth seat cut
– width of ceiling joist
– thickness of temporary sheathing
– ridge board width

Formula for figuring temporary ridge board support height

Step 4–Set Common Rafters

1. Lay out ridge board. If you do not lay out ridge before you put it up, you will need to string a line between the two walls and plumb up to the ridge board.

2. Set two opposing rafters at one end.

3. Set two opposing rafters at the other end.

4. String a line on the top of the ridge board to keep it straight as you set the balance of the rafters.

5. Set the remaining rafters.

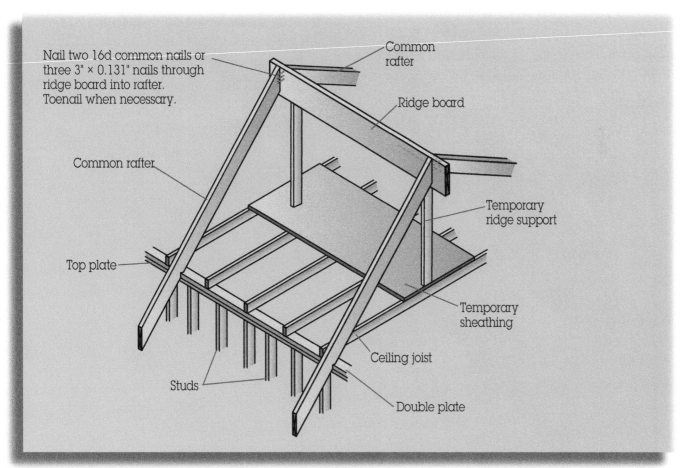

Nail two 16d common nails or three 3" × 0.131" nails through ridge board into rafter. Toenail when necessary.

Common rafter

Ridge board

Common rafter

Temporary ridge support

Top plate

Temporary sheathing

Studs

Ceiling joist

Double plate

Here two pair of common rafters are fitted into place to secure the ridge board. Be sure to set end rafters first.

Framer-Friendly Tips

Laying out the ridge board before putting it up saves time.

Step 5-Find Length of Hip and Valley Rafters

The length of the hip and valley rafters can be found by using any of the six common rafter methods previously described, and then making adjustments for the run and the top and bottom cuts.

Adjustment for the Run

For every 12" of common rafter run, there is 16.97" (17" approx.) of run for hip and valley rafters.

Multiply the run in feet of the common rafter by 16.97" (17" is commonly used) to get the run of the hip or valley rafter.

Adjustments for the Top and Bottom Cuts

The cut mark will be made similar to the common rafter cut mark, except that the hip-val scale on the speed square will be used instead of the common scale to mark the line to cut. (See "Rafter Cut Length" and "Angle Cuts" earlier in this chapter)

If a framing square is used, apply the same procedure shown previously, except use 17" instead of 12" along the blade of the framing square.

These procedures assume a hip or valley corner of 90 degrees.

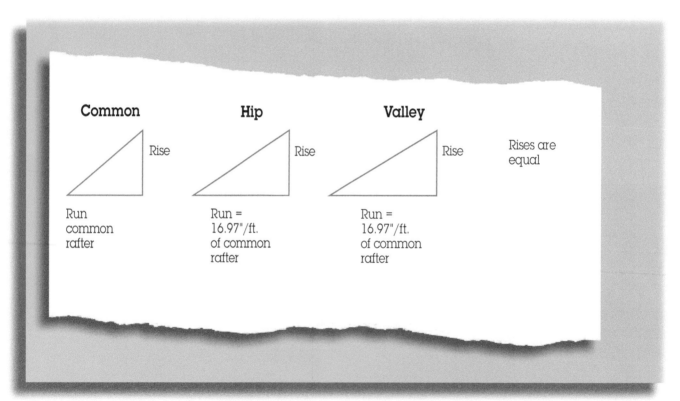

The relationship between common rafters and hip/valley rafters

All of the hip and valley rafters presented in this chapter are based on 90 degree corners. Hip and valley rafters for different angled corners can be found using similar methods to the 90 degree corner. They are still based on the relationship of the rise to the run.

Step 6-Cut Hip and Valley Rafters

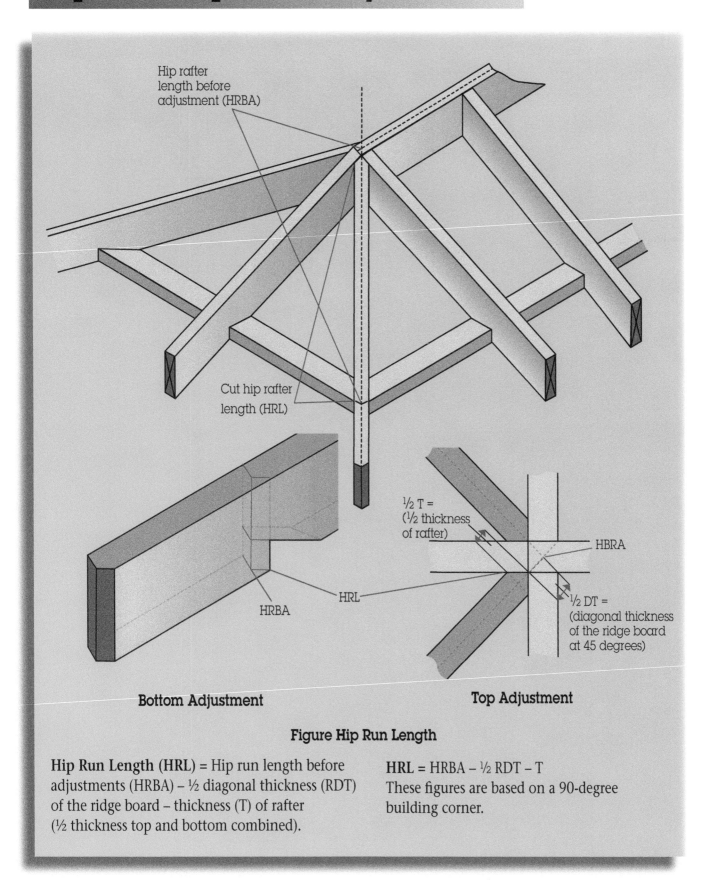

Hip rafter length before adjustment (HRBA)

Cut hip rafter length (HRL)

$\frac{1}{2}$ T = ($\frac{1}{2}$ thickness of rafter)

HBRA

$\frac{1}{2}$ DT = (diagonal thickness of the ridge board at 45 degrees)

HRBA

HRL

Bottom Adjustment

Top Adjustment

Figure Hip Run Length

Hip Run Length (HRL) = Hip run length before adjustments (HRBA) – $\frac{1}{2}$ diagonal thickness (RDT) of the ridge board – thickness (T) of rafter ($\frac{1}{2}$ thickness top and bottom combined).

$HRL = HRBA - \frac{1}{2} RDT - T$
These figures are based on a 90-degree building corner.

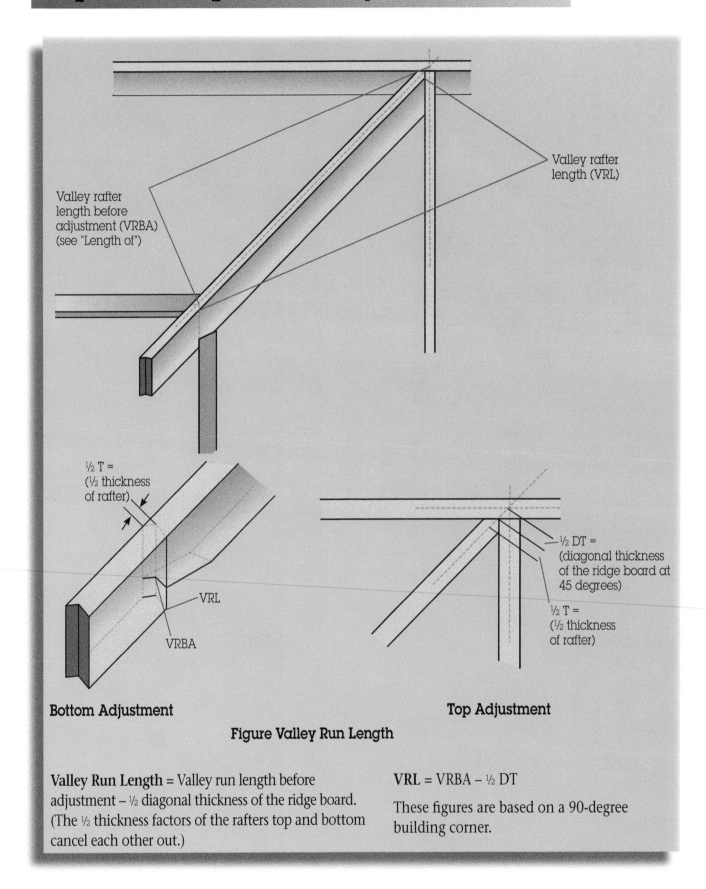

Valley rafter length (VRL)

Valley rafter length before adjustment (VRBA) (see "Length of")

½ T = (½ thickness of rafter)

VRL

VRBA

½ DT = (diagonal thickness of the ridge board at 45 degrees)

½ T = (½ thickness of rafter)

Bottom Adjustment

Top Adjustment

Figure Valley Run Length

Valley Run Length = Valley run length before adjustment – ½ diagonal thickness of the ridge board. (The ½ thickness factors of the rafters top and bottom cancel each other out.)

$$VRL = VRBA - \tfrac{1}{2}\,DT$$

These figures are based on a 90-degree building corner.

Step 7-Set Hip and Valley Rafters

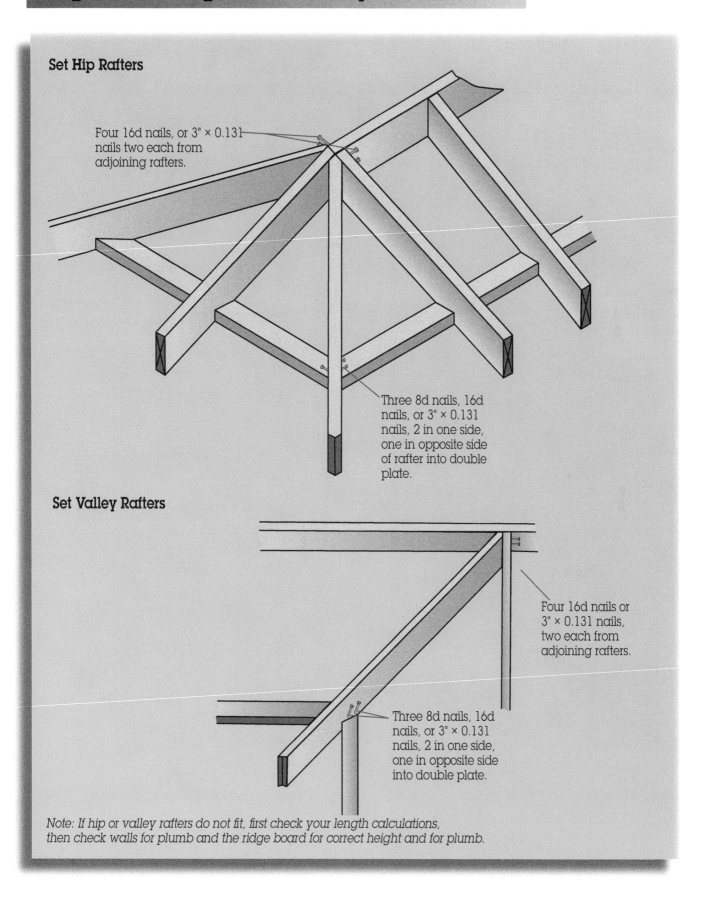

Set Hip Rafters

Four 16d nails, or 3" × 0.131 nails two each from adjoining rafters.

Three 8d nails, 16d nails, or 3" × 0.131 nails, 2 in one side, one in opposite side of rafter into double plate.

Set Valley Rafters

Four 16d nails or 3" × 0.131 nails, two each from adjoining rafters.

Three 8d nails, 16d nails, or 3" × 0.131 nails, 2 in one side, one in opposite side into double plate.

Note: If hip or valley rafters do not fit, first check your length calculations, then check walls for plumb and the ridge board for correct height and for plumb.

Step 8-Set Jack Rafters

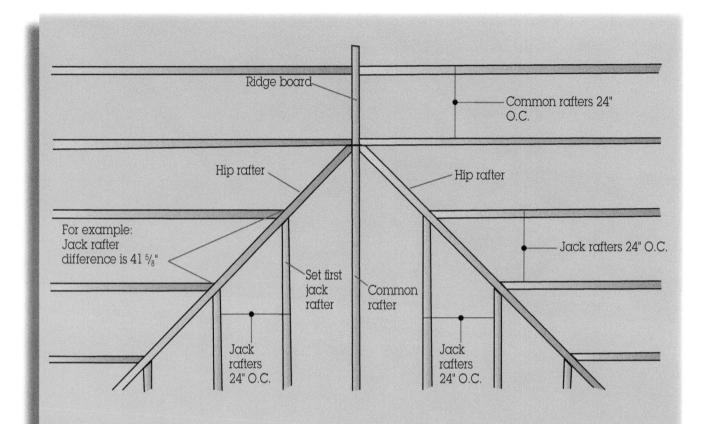

Ridge board

Common rafters 24" O.C.

Hip rafter

Hip rafter

For example:
Jack rafter
difference is 41 ⅝"

Jack rafters 24" O.C.

Set first jack rafter

Common rafter

Jack rafters 24" O.C.

Jack rafters 24" O.C.

Set first jack rafter on 16" or 24" spacing with common rafters.

Measure length from common rafter to first jack rafter and then use standard jack rafter differences, as given in framing square table, to measure lengths of remaining jack rafters along the hip rafter.

For 17 $\frac{12}{|}$ pitch

Framing Square Segment

		23	22	21	20	19	18	17	16
LENGTH	COMMON	RAFTERS	PER FOOT	RUN		21 63	20 81	20
"	HIP OR	VALLEY	"	"	"	24 74	24 02	23 32
DIFF	IN LENGTH	OF JACKS	16 INCHES	CENTERS		28 ⅞	27 ¼	26 ¹¹/₁₆
"	"	"	2 FEET	"		43 ¼	41 ⅝	40
SIDE	CUT	OF	JACKS	USE		6 ¹¹/₁₆	6 ¹⁵/₁₆	7 ³/₁₆
"	"	HIP OR	VALLEY	"		8 ¼	8 ½	8 ¾
	22	21	20	19	18	17	16	15	14

Step 9-Block Rafters and Lookouts
Step 10-Set Fascia

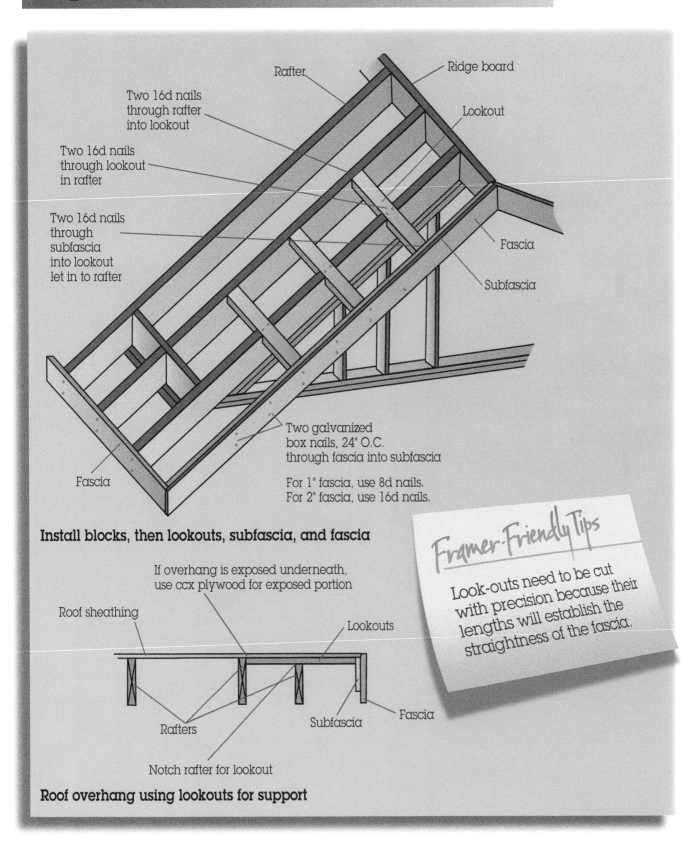

Rafter

Ridge board

Two 16d nails through rafter into lookout

Lookout

Two 16d nails through lookout in rafter

Two 16d nails through subfascia into lookout let in to rafter

Fascia

Subfascia

Fascia

Two galvanized box nails, 24" O.C. through fascia into subfascia

For 1" fascia, use 8d nails.
For 2" fascia, use 16d nails.

Install blocks, then lookouts, subfascia, and fascia

Framer-Friendly Tips

Look-outs need to be cut with precision because their lengths will establish the straightness of the fascia.

If overhang is exposed underneath, use ccx plywood for exposed portion

Roof sheathing

Lookouts

Rafters

Subfascia

Fascia

Notch rafter for lookout

Roof overhang using lookouts for support

Step 11-Install Sheathing

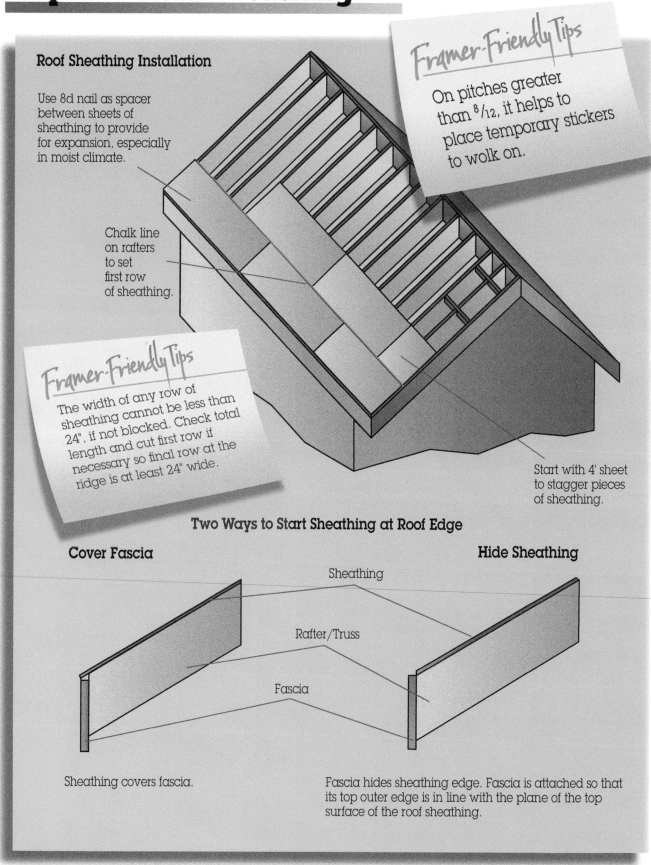

Roof Sheathing Installation

Use 8d nail as spacer between sheets of sheathing to provide for expansion, especially in moist climate.

Chalk line on rafters to set first row of sheathing.

Framer-Friendly Tips

On pitches greater than $^8/_{12}$, it helps to place temporary stickers to wolk on.

Framer-Friendly Tips

The width of any row of sheathing cannot be less than 24", if not blocked. Check total length and cut first row if necessary so final row at the ridge is at least 24" wide.

Start with 4' sheet to stagger pieces of sheathing.

Two Ways to Start Sheathing at Roof Edge

Cover Fascia

Hide Sheathing

Sheathing

Rafter/Truss

Fascia

Sheathing covers fascia.

Fascia hides sheathing edge. Fascia is attached so that its top outer edge is in line with the plane of the top surface of the roof sheathing.

Learning Rafters by Example

Learning by Example

The best way to learn cutting of rafters is to completely work through the actual process. The following 9 examples use the diagonal percent method of finding rafter lengths. The examples show how to find the lengths of the parts of a relatively difficult roof. If you are able to work through and understand the processes, you should be able to figure out how to cut and stack rafters.

Calculators

There are calculators that are made specifically for assisting with construction math. These are very helpful in finding rafter lengths. Construction Master IV® is one available calculator, which we will refer to and use in this chapter to demonstrate the process of finding rafter lengths. These calculators make it easy to do the complicated math, working in feet and inches. The sequence of buttons takes a little time to master, but once you are familiar with them, you will never go back to pencil and paper.

Before we look at the individual examples there will be a brief review and organization of the important parts of rafter cutting.

Considerations for Cutting Rafters

When cutting rafters, you need to consider the following four factors:

1. Figuring rafter length
2. Figuring the adjustment to rafter length at top and bottom
3. Finding the angle cuts at the top, bottom, and at the bird's mouth
4. Finding the bird's mouth height

Construction calculator designed to assist in construction math.

1. Figuring Rafter Length

Figuring Rafter Length Using Diagonal Percent was shown earlier in the chapter with the six ways to figure rafter lengths. (See "F.") When you use the diagonal percent, the most difficult part of figuring rafter length is finding the length of the run. The adjustments that need to be made at the top and bottom of the rafter should be added and subtracted from the run before the rafter length is calculated. In finding the run, it is best to start with the full run distance from the outside of the bearing wall to the framing point of any connecting framing member. Use the framing point for consistency, and then make adjustments from there.

2. Figure the Adjustments to Rafter Length at Top and Bottom

Because there are so many different types of connections for rafters, it helps to establish certain standard ways to connect, and measure them in order to find the proper adjustments to length for the top and bottom. Following are some standard connections and their adjustments. They will not apply to every situation, but they will work for the most common roofs. Rafter cut lengths are figured to outside edge of rafters.

Adjustments for Common Rafters

1. Subtract half the thickness of the ridge board at the top.

2. At the bottom measure to the outside of the wall framing (not the sheathing).

Adjustments for Hip and Valley Rafters

1. At the top, subtract half the 45° thickness of the ridge board. (See "Connection #1" [close-up] illustration later in this chapter and subtract half the thickness of the rafter.)

2. At the bottom, measure to the outside corner of the two connecting walls. Then, for the hip run, subtract half the thickness of the rafter and for the valley run add half the thickness of the rafter.

Adjustments for Jack Rafters

1. Measure to the framing point where it meets the hip or valley it is connected to. (See "Adjusting the Top Length for Jack Rafters" illustration later in this chapter.)

2. Subtract half the 45° thickness of the valley rafter and add half the thickness of the rafter.

3. When the rafter rests on an exterior wall, measure to the outside of the wall framing.

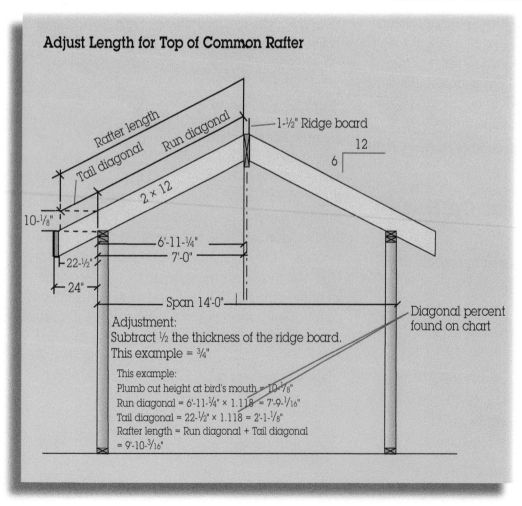

Adjust Length for Top of Common Rafter

Rafter length
Tail diagonal
Run diagonal
1-½" Ridge board
2 × 12
12
6
10-⅛"
6'-11-¼"
7'-0"
22-½"
24"
Span 14'-0"

Adjustment:
Subtract ½ the thickness of the ridge board.
This example = ¾"

This example:
Plumb cut height at bird's mouth = 10-⅛"
Run diagonal = 6'-11-¼" × 1.118 = 7'-9-1/16"
Tail diagonal = 22-½" × 1.118 = 2'-1-⅛"
Rafter length = Run diagonal + Tail diagonal
= 9'-10-3/16"

Diagonal percent found on chart

Adjustments for Rafters Running Between Hips and Valleys

1. Measure to the framing points where it meets the hip or valley it is connecting to. (See "Adjusting the Top Length for Jack Rafters" illustration later in this chapter.)

Adjustments for Miscellaneous Connections Between Hips, Valleys, Ridges, and Rafters

1. Find the combination of cuts that provides the greatest number of standard cuts and still provides a sound structural connection.

2. Measure to the framing point for making adjustments.

3. Connection #2 is an example of a miscellaneous connection where a ridge board, a common rafter, a hip rafter, and a valley rafter connect. (See "Connection #2" illustration later in chapter.)

3. Finding the Angle Cuts at the Top, Bottom, and at the Bird's Mouth

It is easy to figure the angle cuts if you break them down into two separate angles. The first is the **pitch angle**, and the second is the **connection angle**. The pitch angle is either a common or a hip/valley. If you use a speed square, you don't even have to calculate it. If you are cutting a rafter that is not a hip or a valley, then use the common scale on a speed square for the pitch of your roof and draw your pitch angle line on the rafter. If you're cutting a hip or valley rafter, then use the hip-val scale on the speed square. The pitch angle line will be your cut line for your saw cut.

The connection angle depends on a lot of factors, but 45° and 90° are the most commonly used angles. Basically, you will be setting the angle of your saw at the connection angle and cutting the cut line created by the pitch angle. For 90° corners on hips and valleys, the connection angle for jack rafters will be 45°. For standard, common rafters, the top connection angle is 90°.

4. Finding the Bird's Mouth Height

The height of the bird's mouth will affect the height of the roof and possibly the interior design of the ceiling. The most common detail for a bird's mouth has the bird's mouth cut starting at the inside corner of the wall.

On a hip and valley, the inside corner won't align with the wall. Since the height of the hip and valley bird's mouth must be the same as the common rafter bird's mouth, you can simply measure the common rafter height and transfer it to the hip and valley bird's mouth. The chart on page 105 shows some common bird's mouth heights.

On hip rafters, you measure the height to the outside edge of the hip, whereas on valley rafters, it's a little tricky. You need to measure to the center of the valley, which is slightly higher than the outside edge.

Finding Rafter Length: 9 Examples

Breaking the process of cutting rafters into the four basic characteristics described in this chapter helps to organize the task, but it is still a complicated process. Probably the best way to learn is to work through the steps in figuring individual rafters. The following illustration is an example of a roof that has 8 different rafters and 1 ridge board identified. The nine examples explain how to find the lengths for these rafters and ridge board based on the illustration.

Finding Common Rafter Length— Example ① on Roof Example Illustration

The roof span at this area is 28'-0", making the run equal to ½ the span of 14'-0" minus half the thickness of the ridge board (¾"). That makes the adjusted run 13'-11-¼". Multiplying that times the diagonal percent for a ⁶/₁₂ pitch roof (which is 1.118) gives a run diagonal length of 15'-7". If you add that length to the tail diagonal of 2'-1-⅛", the rafter length is 17'-8-⅛". The tail diagonal

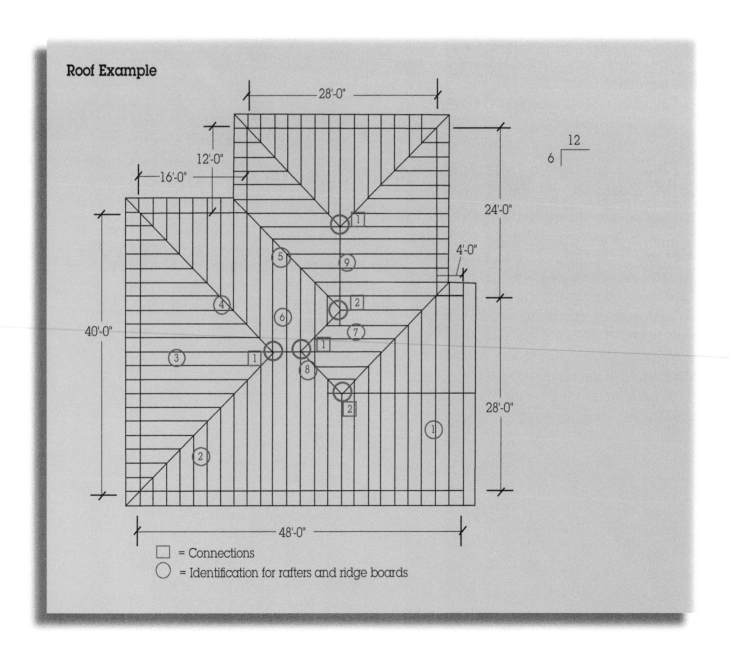

Roof Example

□ = Connections

○ = Identification for rafters and ridge boards

is found by subtracting the fascia (1-½") from the 2'-0" overhang, which gives (22-½"), and multiplying by the diagonal percent 1.118.

Finding a Jack Rafter Length— Example ② on Roof Example Illustration

Most hip rafters are on 90° corners, with the hip in the middle of the corner. Because the two sides of a triangle made by a 90° angle and two 45° angles are the same, the run of the jack rafter can be easily found.

The distance of your layout to the center of your rafter is the same distance as your run to the center of your hip. Just subtract one half the thickness of the hip at a 45° angle (1-¹/₁₆") from the run and add half the thickness of the rafter to the run. Then add on your tail diagonal length. This will give you your rafter length. In this example, the rafter is on layout at 8'-0", so we subtract 1-¹/₁₆" (half the thickness of the 1-½" hip at 45°) and add half the thickness of the rafter (¾" for a 1-½" rafter), giving 7'-11-¹¹/₁₆", which is multiplied by the diagonal percent of 1.118. The result is 8'-11". Add this to the tail diagonal of 2'-1-⅛" (same as common rafter tail), and we get a jack rafter length of 11'-⅛".

Finding a Ridge End Common Rafter Length—Example ③ on Roof Example Illustration

As long as you use the top cut illustration in "Connection #1," then this rafter will be cut the same length as the king common rafter adjacent to it.

Finding a Hip Length—Example ④ on Roof Example Illustration

Finding the hip length requires an additional step and uses the hip-val diagonal percent. First find the hip run. It is the diagonal created by a triangle in which the other two sides are the run of the ridge end common and the line from the hip corner to the ridge end common. In this case, the span is 40', so the run is 20', and the distance from the corner is also 20'. Using the calculator, enter 20' for the run, 20' for the rise, and press the diagonal button. The result is 28'- 3-⁷/₁₆". This distance is the run of your hip to the exterior wall. Then make top adjustment and add the tail. To make the top adjustment subtract half the distance of the ridge at a 45° angle, which for a 1-½" ridge is 1-¹/₁₆". Then subtract half the thickness of the rafter to adjust for the top 45° V cut, which for a 1-½" hip would be ¾". This leaves a top adjusted hip run of 28'- 1-⅝".

Then find the hip tail length using a similar procedure. The sides are 2', which leads to a 2'-9-¹⁵/₁₆" diagonal. Then subtract 1-½" at a 45° angle for the fascia (which is 2-⅛") and subtract half the thickness of the rafter to adjust for the bottom 45° V cut, which for a 1-½" hip rafter would be ¾", so the hip tail run is 2'-7-¹/₁₆". Add this figure to the 28'- 1-⅝" hip run, and you get a hip rafter run of 30'-8-¹¹/₁₆". Multiplying that number by the hip-val diagonal percent of 1.061 results in a hip rafter length of 32'-7-³/₁₆". Remember, these lengths are to the edge of the rafter. The cuts are 45° angle cuts at a ⁶/₁₂ hip-val pitch angle.

Finding a Valley Rafter Length— Example ⑤ on Roof Example Illustration

This valley will be the same length as the hip rafter for the 28'-0" span section, except for the end cuts. On the bottom, the 45° cuts will be concave (<) instead of convex (>) like the hip. At the top because the ridge board goes on up from the end of the valley, there will be a full-width 45° cut. The top adjustment will require you to subtract one half the thickness of the ridge at 45°, which is 1-¹/₁₆". Then add half of the rafter thickness, which for a 1-½" rafter is ¾". This will be to the long side of the cheek cut.

Connection # 1

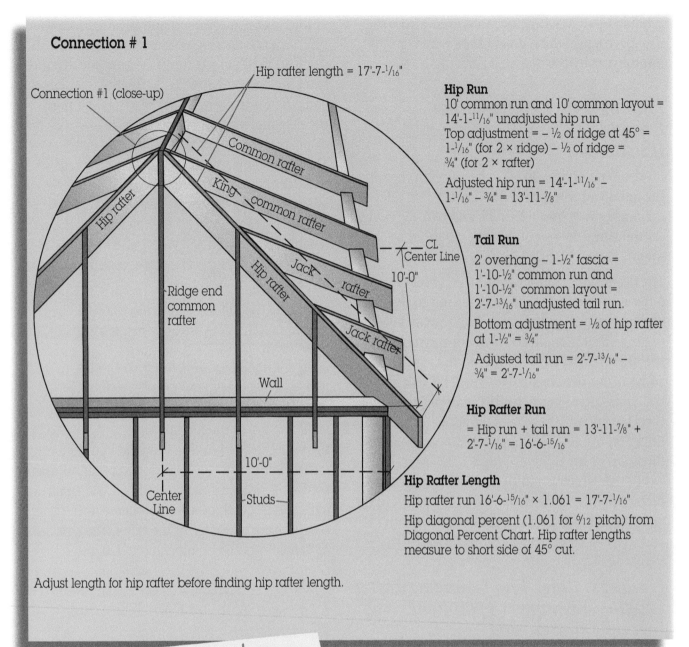

Connection #1 (close-up)

Hip rafter length = 17'-7-1/16"

Common rafter

King common rafter

Hip rafter

Jack rafter

Hip rafter

Jack rafter

Ridge end common rafter

CL Center Line

10'-0"

Wall

Center Line

10'-0"

Studs

Hip Run

10' common run and 10' common layout = 14'-1-11/16" unadjusted hip run

Top adjustment = – 1/2 of ridge at 45° = 1-1/16" (for 2 × ridge) – 1/2 of ridge = 3/4" (for 2 × rafter)

Adjusted hip run = 14'-1-11/16" – 1-1/16" – 3/4" = 13'-11-7/8"

Tail Run

2' overhang – 1-1/2" fascia = 1'-10-1/2" common run and 1'-10-1/2" common layout = 2'-7-13/16" unadjusted tail run.

Bottom adjustment = 1/2 of hip rafter at 1-1/2" = 3/4"

Adjusted tail run = 2'-7-13/16" – 3/4" = 2'-7-1/16"

Hip Rafter Run

= Hip run + tail run = 13'-11-7/8" + 2'-7-1/16" = 16'-6-15/16"

Hip Rafter Length

Hip rafter run 16'-6-15/16" × 1.061 = 17'-7-1/16"

Hip diagonal percent (1.061 for 6/12 pitch) from Diagonal Percent Chart. Hip rafter lengths measure to short side of 45° cut.

Adjust length for hip rafter before finding hip rafter length.

Framer-Friendly Tips

When using I-joist rafters, the cut bottom flange must rest entirely on the wall.

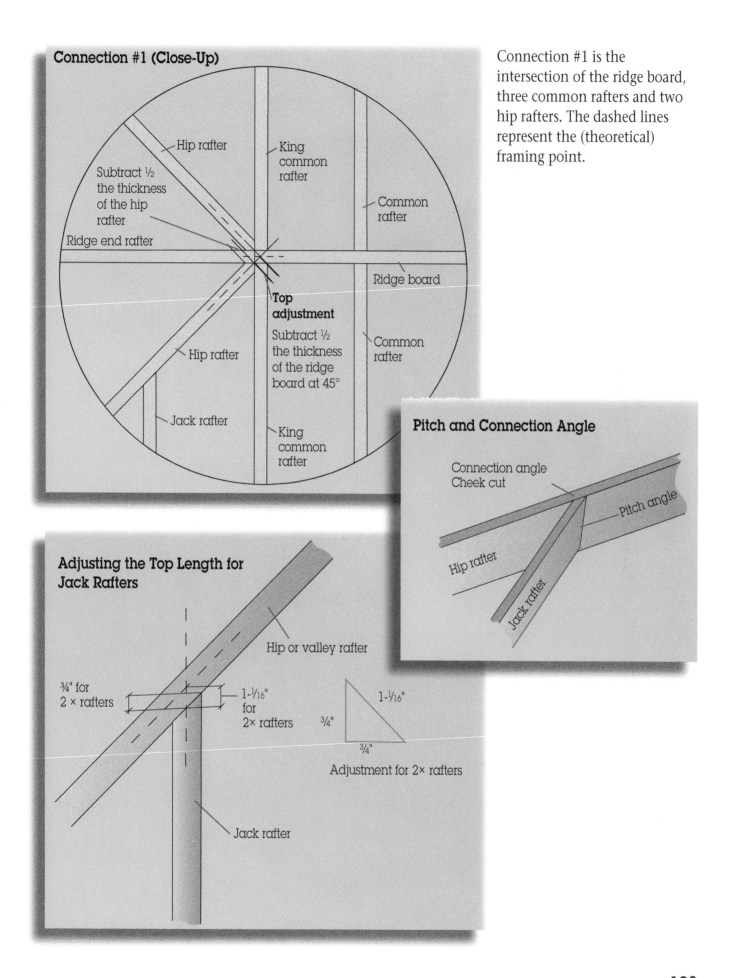

Connection #1 (Close-Up)

Hip rafter

Subtract ½ the thickness of the hip rafter

Ridge end rafter

King common rafter

Common rafter

Common rafter

Ridge board

Top adjustment

Subtract ½ the thickness of the ridge board at 45°

Hip rafter

Jack rafter

King common rafter

Connection #1 is the intersection of the ridge board, three common rafters and two hip rafters. The dashed lines represent the (theoretical) framing point.

Pitch and Connection Angle

Connection angle
Cheek cut

Pitch angle

Hip rafter

Jack rafter

Adjusting the Top Length for Jack Rafters

Hip or valley rafter

¾" for 2 × rafters

1-1/16" for 2× rafters

1-1/16"

¾"

¾"

Adjustment for 2× rafters

Jack rafter

103

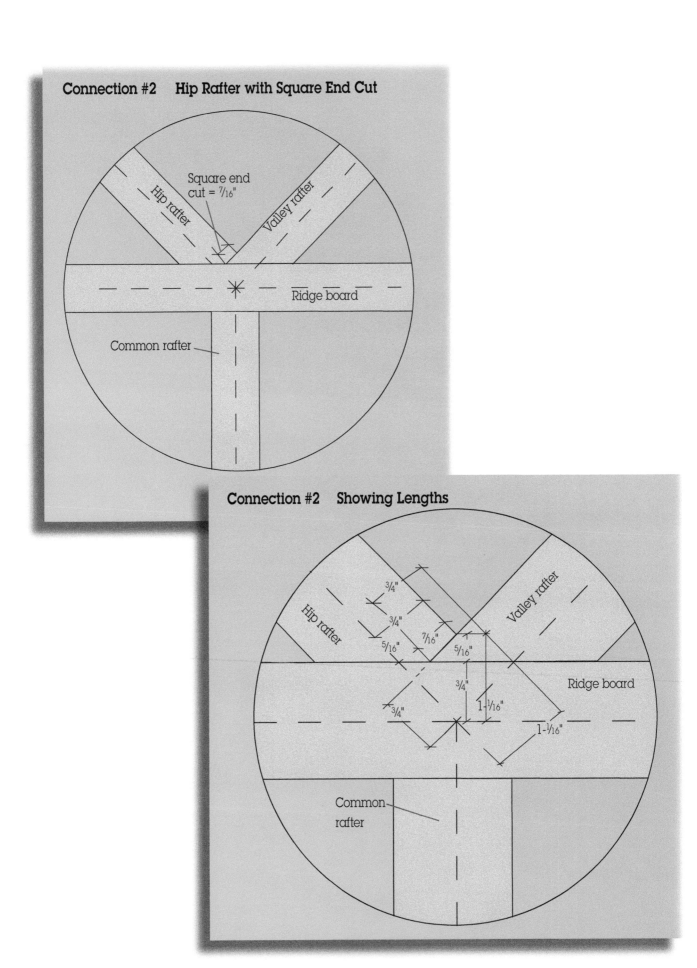

Connection #2   Hip Rafter with Square End Cut

Square end
cut = 7/16"

Hip rafter

Valley rafter

Ridge board

Common rafter

Connection #2   Showing Lengths

3/4"

3/4"

5/16"

7/16"

5/16"

Hip rafter

Valley rafter

3/4"

1-1/16"

3/4"

Ridge board

1-1/16"

Common
rafter

Finding the valley rafter length is similar to finding the hip length, and requires the following steps:

- Span = 28'-0"

- Run = 14'-0"

- Find valley run. On the calculator enter 14' and press the Rise button, then enter 14' and press the Run button and then press the Diag button. = 19'-9-9/16"

- Top adjustment = subtract ½ ridge board at 45° = 1-1/16" and add ½ the thickness of the

valley rafter = ¾". This is to the long side. See Connection #2 illustration

- Adjusted valley run = 19'-9-9/16" – 1-1/16" + ¾" = 19'-9-¼"

- Find valley tail run. On the calculator enter 2' (2'-0" overhang) and press the Rise button, then enter 2' and press the Run button and then press the Diag button. = 2'-9-15/16".

- Bottom adjustment = subtract 2-⅛" (1-½" fascia at 45°) and add ½" thickness of the valley rafter = ¾" for V cut.

  - Adjusted tail run = 2'-9-15/16" – 2-⅛" + ¾" = 2'-8-9/16"

  - Add the adjusted valley run and the adjusted tail valley run = 19'-9-¼" + 2'-8-9/16" = 22'-5-13/16"

  - Valley rafter length = 22'-5-13/16" × 1.061 (hip-val diagonal percent) = 23'-10-¼"

The top will be a 45° saw cut for the connection angle at a 6/12 hip-val cut for the pitch angle.

The bottom will be concave (<), two 45° saw cuts at a 6/12 hip-val cut.

Finding a Valley-to-Ridge Jack Rafter—Example ⑥ on Roof Example Illustration

There are a couple of ways to find the length of this rafter. The ridge location is easy to establish as half the span of 40', making it 20'. The valley point can be determined by figuring the distance the valley runs before the rafter starts. In this case, since the rafters all conveniently line up and run at 24" O.C., the easiest method is to count the rafter spaces from the other side of the roof. In this example, there are seven rafter spaces; therefore the run will be 14'.

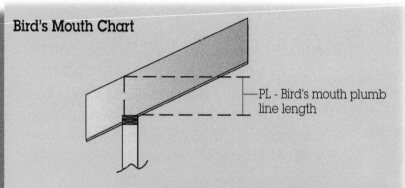

Bird's Mouth Chart

PL - Bird's mouth plumb line length

Bird's Mouth Plumb Line Lengths
2 × 6 Walls

Rafter Size	11¼	9¼	7¼	5½
Pitch				
3/12	10¼	8 3/16	6⅛	4 5/16
4/12	10 1/16	7 15/16	5 13/16	4
5/12	9⅞	7 11/16	5 9/16	3 5/8
6/12	9 13/16	7 9/16	5 3/8	3 3/8
7/12	9 13/16	7½	5 3/16	3 3/16
8/12	9 13/16	7 9/16	5	2 15/16
9/12	9 15/16	7 9/16	4 15/16	2 3/4
10/12	10 1/16	7½	4 7/8	2 5/8
11/12	10 3/16	7½	4 3/4	2 3/8
12/12	10 7/16	7 9/16	4 3/4	2 1/4

2 × 4 Walls

Rafter Size	11¼	9¼	7¼	5½
Pitch				
3/12	10 ¾	8 11/16	6 5/8	4 13/16
4/12	10 11/16	8 9/16	6 7/16	4 5/8
5/12	10 ¾	8 9/16	6 7/16	4 ½
6/12	10 13/16	8 9/16	6 3/8	4 3/8
7/12	10 15/16	8 5/8	6 5/16	4 5/16
8/12	11 3/16	8 13/16	6 3/8	4 5/16
9/12	11 7/16	8 15/16	6 7/16	4 ¼
10/12	11 11/16	9 1/8	6 ½	4 ¼
11/12	12 1/16	9 5/16	6 5/8	4 ¼
12/12	12 7/16	9 9/16	6 3/4	4 ¼

Charts provide bird's mouth plumb line lengths.

Subtract half the distance at 45° for the valley rafter (1-1/16"), and add half the thickness of the rafter (3/4") for the bottom adjustment and subtract half the thickness of the ridge board (3/4") at the top adjustment. The run will be 13'-10-15/16". The rafter length will be 13'-10-15/16" × 1.118 (diagonal percent) resulting in a 15'-6-5/8" rafter length. The connection angle at the top will be a 90° saw cut, and the pitch will be at a 6/12 common cut on the speed square. The bottom will be a 45° saw cut at a 6/12 common cut. The measurement will be to the long point of the 45° cheek cut.

Finding Valley-to-Hip Jack Rafter Length—Example ⑦ on Roof Example Illustration

There are different ways to find the run length. Here is a way that has not yet been illustrated. In this example, run length will be figured from the 28' span length. The run for the 28' span is 14'. The

Jack rafter run lengths equal layout lengths.

top of the rafter is 2' past the end of the ridge board, which will add 2' to the run going up the hip that it connects to.

The run at the bottom will be shortened by 4' because it extends up the valley the equivalent of 4' of run. This leaves 12' of run. Adjust for top and bottom by subtracting one half of a 45° angle for top and bottom cuts or two times 1-1/16" (2-1/8") = 11'-9-7/8" times 1.118. This makes for a rafter length of 13'-2-5/8". Both the top and bottom would have a 45° cheek cut for the connection angle and would be marked at a common 6/12 for the pitch angle.

Ridge-to-Ridge Hip Rafter—Example ⑧ on Roof Example Illustration

In this example, the rafters are so conveniently arranged that we can see the hip rafter goes from the center of one rafter to the center of another rafter with two in between, resulting in a distance of 6'.

Another way to find this length is to calculate the difference in the runs for the ridges that establish the height difference. One has a span of 28'-0" for a run of 14'-0", while the other has a span of 40'-0" for a run of 20'-0". The difference is 6'-0", the same as we just figured.

Once you have the 6'-0" of run, then you follow the same procedure as with a hip and make the necessary top and bottom adjustments. First establish the hip run. Enter 6'-0" run and 6'-0" rise on the calculator and press diagonal, which gives you the hip run of 8'-5-13/16". The top will be a standard hip connection. Therefore one half the ridge at a 45° angle (1-1/16") will be subtracted. At the bottom it will be a #2 connection. (Connection #2.) Therefore subtract one half the thickness of the ridge at a 45° angle, or 1-1/16". The bottom will also require a square cut $\frac{7}{16}$" deep on the end. You can establish the thickness of this square cut by finding the diagonal for the triangle in which the other two sides are the same and created by the balance of the difference between half the distance of the ridge board at 45° and half the distance of the ridge board at 90°.

The $\frac{7}{16}$" square cut will not affect your hip run. This means that you can subtract the 1-$\frac{1}{16}$" and 1-$\frac{1}{16}$" to get an adjusted hip run of 8'-3-$\frac{11}{16}$". Multiplying 8'-3-$\frac{11}{16}$" × 1.061 (hip-val diagonal percent) gives you the ridge to ridge hip rafter length of 8'-9-¾". The top cut will be a regular hip cut with convex (>) 45° cuts at a $\frac{6}{12}$ hip-val pitch. The bottom will be a 45° cut at a $\frac{6}{12}$ hip-val pitch with a $\frac{7}{16}$" square cut end.

Finding the Ridge Board Length— Example ⑨ on Roof Example Illustration

The ridge board runs parallel with the wall at the other end of valley #5 and the hip of the 28'-0" span. That length is 12'-0", so the length of the ridge is 12'-0" with adjustments at the ends. The hip connection is a number 1, so one half of the thickness of the common rafter (¾") is added. At the other end, it is a connection #2, and the ridge will extend to the next rafter, adding 23-¼" to the length.

The ridge board length therefore is: 12'-0" + ¾" + 23-¼" = 14'-0". Both ends will be cut at 90° with square ends.

Summary

Until you have framed many roofs, cutting rafters is always going to be a challenge. Three ways to make it easier are:

- First, use the diagonal percent to find the rafter length.

- Second, figure lengths to the framing points and then make the adjustments.

- Third, become familiar with and use a construction calculator for the math.

If ever you get stumped, you can always organize your thinking by using the four basic characteristics of cutting rafters:

1. Find the length.

2. Adjust for the top and bottom.

3. Figure the angle cuts for the top, bottom, and bird's mouth.

4. Figure the height of the bird's mouth.

Framer-Friendly Tips

Sketching up what is unclear on a scrap piece of paper can help you to understand the connection better.

Ceiling Joists

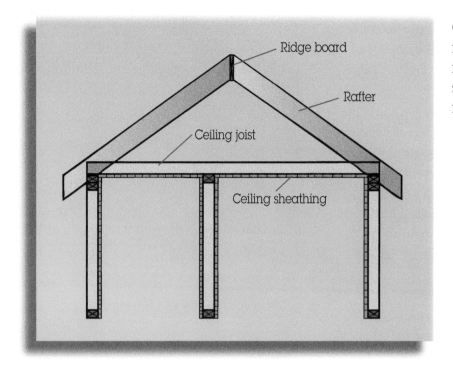

Ceiling joists serve two basic functions: they provide support for the ceiling sheathing and the support needed to keep the rafters from pushing the walls out.

Rafters that are not supported at the top or somewhere along the span by a beam create an outward force on the wall they rest on. This force is frequently offset by ceiling joists joining the walls on the opposite side of the roof.

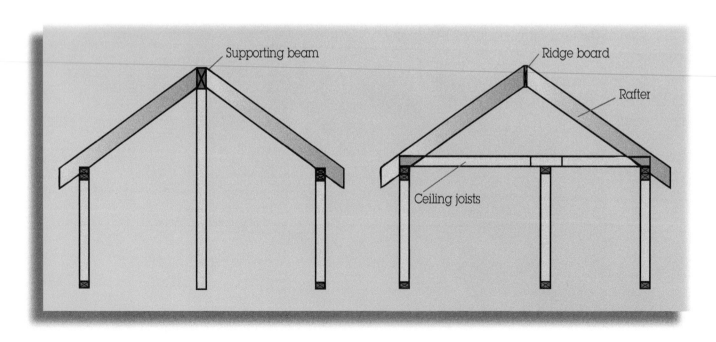

Ceiling Joists (continued)

The chart below shows a common size of ceiling joist needed for different spans.

Ceiling Joist Span Chart

24" O.C. 20 PSF uninhabitable

Species and Grade	Maximum Ceiling Joist Spans		
	2 × 4	2 × 6	2 × 8
Douglas Fir-Larch # 1	7'-8"	11'-2"	14'-2"
Douglas Fir-Larch # 2	7'-2"	10'-6"	13'-3"
Douglas Fir-Larch # 3	5'-5"	7'-11"	10'-0"
Hem-Fir # 1	7'-6"	10'-11"	13'-10"
Hem-Fir # 2	7'-1"	10'-4"	13'-1"
Hem-Fir # 3	5'-5"	7'-11"	10'-0"
Southern Pine # 1	8'-0"	12'-6"	15'-10"
Southern Pine # 2	7'-8"	11'-0"	14'-2"
Southern Pine # 3	5'-9"	8'-6"	10'-10"
Spruce-Pine-Fir # 1	7'-2"	10'-6"	13'-3"
Spruce-Pine-Fir # 2	7'-2"	10'-6"	13'-3"
Spruce-Pine-Fir # 3	5'-5"	7'-11"	10'-0"

The bottom chord of roof trusses acts as a ceiling joist and provides support for the ceiling sheathing, and support to keep the walls from pushing out.

Attic areas above ceiling joists must be made accessible if there is a clear height of 30" or more.

This requires framing an attic opening. The opening must not be less than 22" by 30" and have at least 30" clear space above. The attic access should be framed similar to an opening in floor joists as illustrated on "Step 5–Frame Openings in Joists" in Chapter 3.

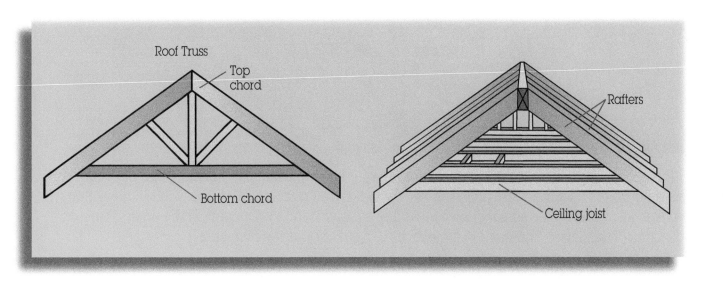

Roof Truss

Top chord

Bottom chord

Rafters

Ceiling joist

Step 1-Spread Trusses

Trusses spread ready to roll

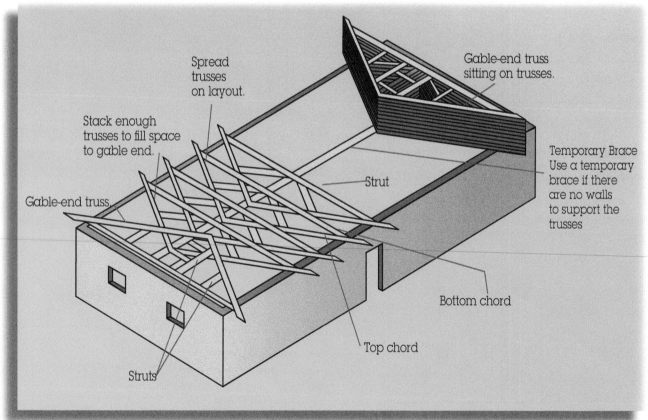

When trusses are delivered in stacks, they should be set on the roof to allow for easy spreading. The gable ends should be on top because they go up first. The direction of the ridge is important so they can be spread and tilted up easily.

When spreading the trusses, place them on your layout marks so that when you roll them, you will have minimum moving of the trusses. They are easier to move lying down.

Steps 5-8

5. Block trusses

8. Sheathing
The skin for the roof, s...
diaphragm.

Step 2-Sheathe Gable Ends

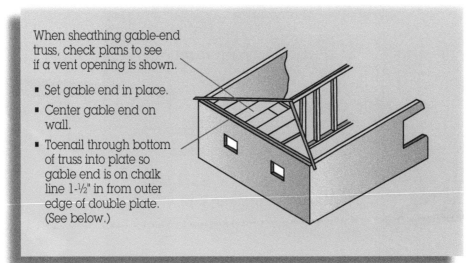

When sheathing gable-end truss, check plans to see if a vent opening is shown.

- Set gable end in place.
- Center gable end on wall.
- Toenail through bottom of truss into plate so gable end is on chalk line 1-½" in from outer edge of double plate. (See below.)

Step 3-Set Gable Ends

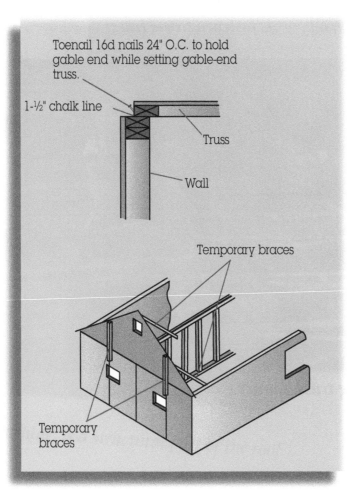

Toenail 16d nails 24" O.C. to hold gable end while setting gable-end truss.

1-½" chalk line

Truss

Wall

Temporary braces

Temporary braces

Step

String li

As each tr

Steps

1. String line
2. Lift single
3. Center trus
4. Nail truss t
5. Nail throug
   near ridge o
   marked on t
6. Set six truss
   for plumb a
   brace on gab
   should conn
   interior wall
   between the t
7. Every eight tr
   brace. Refer to
   additional bra

Chapter Six
DOORS, WINDOWS, AND STAIRS

Contents

Chapter Six

DOORS, WINDOWS, AND STAIRS

Doors and windows are two of the few finish items that framers sometimes handle. It is important that time and care are taken to ensure they are installed in a proper, professional manner. Put your framing hammer in the toolbox and use, instead, a lighter, smoother-faced trim hammer and a nail set.

Exterior, pre-hung doors are the type covered in this chapter. They are the ones framers most commonly work with, and most of the skills involved in hanging them will carry over to the hanging of interior doors. The first door you hang on any job will give you the most difficulty. If you have more than one door to hang, do them one after the other; each door will go in a little easier than the one before.

Nail-flange windows and sliding glass doors will vary depending on the manufacturer. You will find here the basic principles of their installation. Use common sense and follow the directions provided, and you should have little trouble installing these units.

Stairs represent one of the more difficult challenges to a framer's skill. As in roof framing, the geometry is a bit complicated, but taken step-by-step, the logic soon becomes clear, leading to successful execution of the plans. There are many different stair designs. The stair layouts described in this chapter are typical. Be aware that the dimensions given on the plans do not always allow for enough headroom. Always check headroom and other dimensions by taking accurate on-site measurements. This chapter also contains instructions for laying out and framing circular stairs.

The instructions for installation of pre-hung doors, windows, sliding glass doors, straight stairs, and circular stairs are all presented in steps for easier understanding. A calculator is handy—some might say necessary—for finding the rise and tread dimensions when not given on the plans. Always double- or triple-check your calculations. Remember: measure twice, cut once. The finish floors at the top and bottom of stairs are often different. When cutting stair stringers, don't forget to check the plans for such differences and then check the height of your top and bottom risers to allow for them.

Door Framing Terms

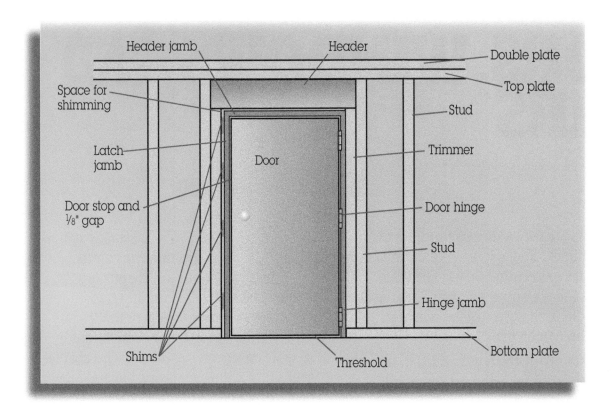

Header jamb · Header · Double plate · Top plate · Space for shimming · Stud · Latch jamb · Trimmer · Door · Door stop and ⅛" gap · Door hinge · Stud · Hinge jamb · Shims · Threshold · Bottom plate

Installation of Exterior Doors

Steps for Pre-Hung Doors

1. Read instructions that come with door.
2. Check plans for door swing.
3. Check threshold for level.
4. Nail hinge jamb.
5. Nail latch jamb using shims.
6. Check door for overall fit.
7. Nail door.
8. Set nails and break or cut shims.
9. Check again for overall fit.

Installation of Exterior Doors (Continued)

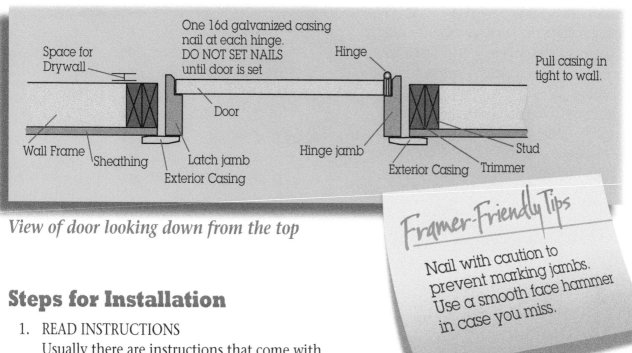

View of door looking down from the top

Steps for Installation

1. **READ INSTRUCTIONS**
   Usually there are instructions that come with the door. Check the instructions over for anything you might need to know.

2. **CHECK PLANS**
   Check the building plans to find the direction of the door swing.

3. **CHECK THE THRESHOLD**
   Check the threshold for level. Shim under hinge jamb if necessary.

4. **NAIL HINGE JAMB**
   Nail hinge jamb tight to trimmer with one 16d galvanized casing nail at each hinge. Do not set nails. Plumb both directions. Shim behind jamb if necessary to obtain plumb or if door needs to be centered in opening.

Framer-Friendly Tips

Nail with caution to prevent marking jambs. Use a smooth face hammer in case you miss.

Framer-Friendly Tips

For a stronger connection remove one screw from each hinge and replace it with a longer screw to go through the jamb, the shims, and into the stud-trimmer.

5. NAIL LATCH JAMB

   Before nailing the latch jamb, make sure:

   • There is a continuous gap between door and jamb of ⅛ at latch jamb and header jamb.

   • Door touches latch jamb equally at top and bottom.

   • Lockset hole in the door and latch jamb line up.

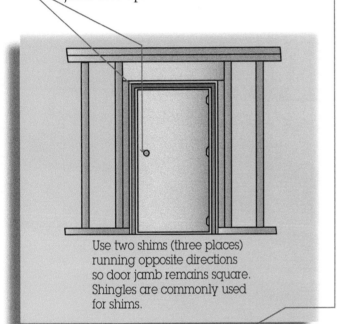

Use two shims (three places) running opposite directions so door jamb remains square. Shingles are commonly used for shims.

6. CHECK DOOR FOR OVERALL FIT:

   • Gap around door is even.

   • Door and latch jamb align.

   • Door closes smoothly, no binding.

   • Lockset holes line up.

   • Door closes tight, no rattle.

7. NAIL DOOR

   • Six 16d galvanized casing nails in hinge jamb—two at each hinge.

   • Four 16d galvanized casing nails in latch jamb—one each top and bottom and two near latch. This includes nails used in Steps 4 and 5.

8. SET NAILS AND TRIM SHIMS

9. FINAL CHECK

   • Check again for overall fit. (See Step 6.)

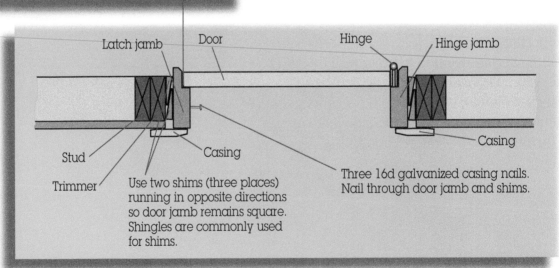

Latch jamb   Door   Hinge   Hinge jamb

Stud

Trimmer

Use two shims (three places) running in opposite directions so door jamb remains square. Shingles are commonly used for shims.

Casing

Casing

Three 16d galvanized casing nails. Nail through door jamb and shims.

View of door looking down from the top

Installation of Nail-Flange Window

1. Install window flashing. (See the following section.)

2. Set window in place.

3. Place temporary shims under bottom of window. Equalize space at top and bottom of window. Shims are usually ⅛" to ¼". Level windowsill.

4. Make gaps the same between window frame and trimmer on each side.

5. Nail top corners from outside.

6. Nail one bottom corner. (Do not set all the way.)

7. Place window slider in and check to see if the gap between the window slider and window frame is the same from top to bottom.

8. If the gap is not equal, check both the rough opening and the window for square and plumb and adjust accordingly.

9. Finish nailing. Make sure gap is equal top, bottom, and middle. (Do not nail top of the window.)

Framer-Friendly Tips

If you're using aluminum nail-flange windows, don't use galvanized nails. They will react chemically or corrode.

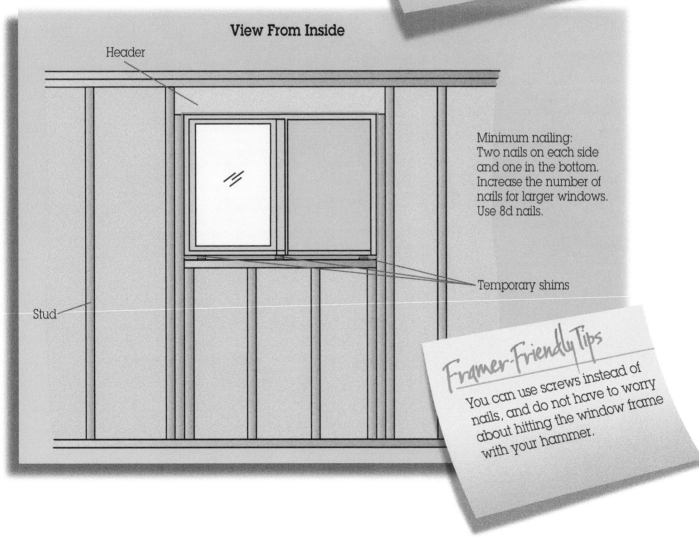

View From Inside

Header

Minimum nailing: Two nails on each side and one in the bottom. Increase the number of nails for larger windows. Use 8d nails.

Temporary shims

Stud

Framer-Friendly Tips

You can use screws instead of nails, and do not have to worry about hitting the window frame with your hammer.

Installation of Window Flashing

Moisture penetration in buildings can cause rot in the structure. New and more extensive exterior rain protective systems have been developed and used to combat this problem. Some of these systems use a special type of water resistant barrier and self-adhesive flashing. Another system is the rainscreen. It provides a whole second layer of protection, typically by installing furring strips over the initial water-resistive barrier and then an outside siding material attached to the furring strips. These furring strips allow for ventilation and pressure equalization. These systems have not been standardized yet and so it is important to follow the specifications as outlined on the plans.

Window flashing is typically installed in one of two methods. Before the weather resistant barrier is put on or after the weather resistant barrier is put on. The following two pages illustrate typical steps used to install flashing before weather resistant barrier is put on and after weather resistant barrier is put on.

If instructions are provided with the windows then be sure to follow the instructions provided, however if not be sure to follow the instructions that follow with the inclusion of the Scot Dam which is meant to provide the required pan flashing.

Steps for Window Flashing Installation
When Windows Installed Before WRB (Weather Resistant Barrier)

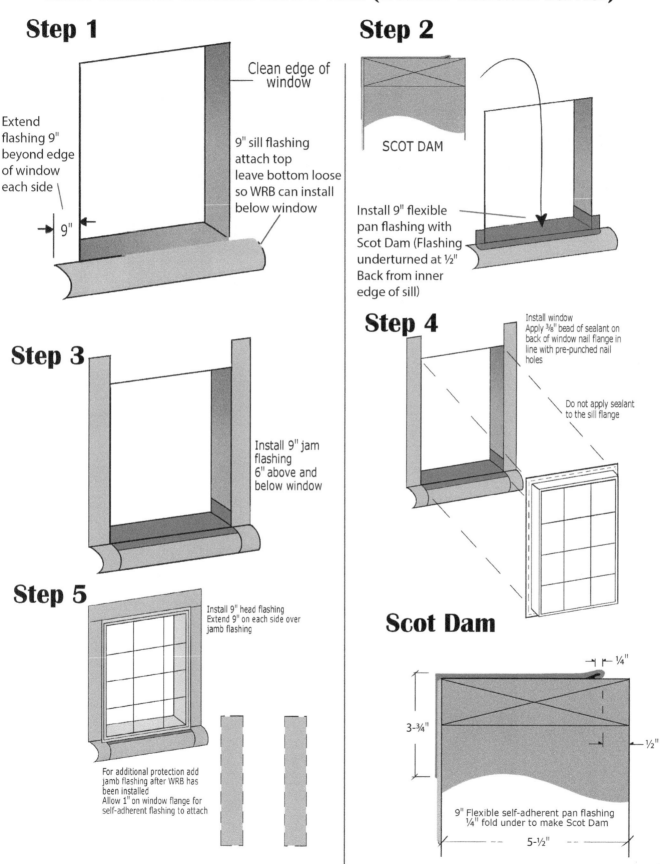

Step 1

Clean edge of window

Extend flashing 9" beyond edge of window each side

9"

9" sill flashing attach top leave bottom loose so WRB can install below window

Step 2

SCOT DAM

Install 9" flexible pan flashing with Scot Dam (Flashing underturned at ½" Back from inner edge of sill)

Step 3

Install 9" jam flashing 6" above and below window

Step 4

Install window
Apply ⅜" bead of sealant on back of window nail flange in line with pre-punched nail holes

Do not apply sealant to the sill flange

Step 5

Install 9" head flashing
Extend 9" on each side over jamb flashing

For additional protection add jamb flashing after WRB has been installed
Allow 1" on window flange for self-adherent flashing to attach

Scot Dam

¼"

3-¾"

½"

5-½"

9" Flexible self-adherent pan flashing ¼" fold under to make Scot Dam

123

Steps for Window Flashing Installation
When Window Installed After WRB (Weather Resistant Barrier)

Step 1

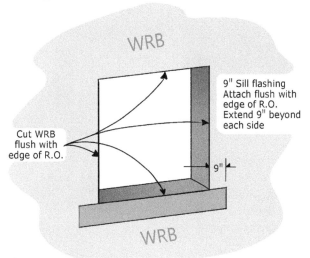

WRB

Cut WRB flush with edge of R.O.

9" Sill flashing Attach flush with edge of R.O. Extend 9" beyond each side

9"

WRB

Step 2

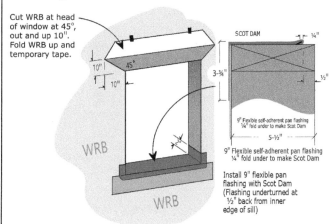

Cut WRB at head of window at 45°, out and up 10". Fold WRB up and temporary tape.

10"
45°
10"

WRB

WRB

SCOT DAM

¼"
3-¾"
½"

9" Flexible self-adherent pan flashing ¼" fold under to make Scot Dam

5-½"

9" Flexible self-adherent pan flashing ¼" fold under to make Scot Dam

Install 9" flexible pan flashing with Scot Dam (Flashing underturned at ½" back from inner edge of sill)

Step 3

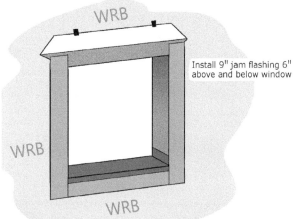

WRB

WRB

WRB

Install 9" jam flashing 6" above and below window

Step 4

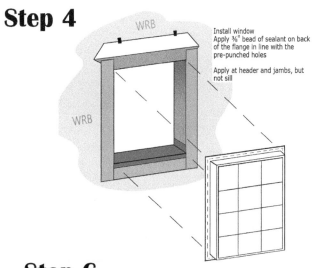

WRB

WRB

Install window Apply ⅜" bead of sealant on back of the flange in line with the pre-punched holes

Apply at header and jambs, but not sill

Step 5

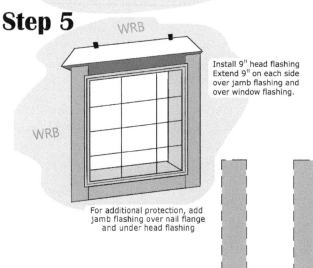

WRB

WRB

Install 9" head flashing Extend 9" on each side over jamb flashing and over window flashing.

For additional protection, add jamb flashing over nail flange and under head flashing

Step 6

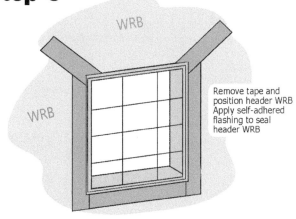

WRB

WRB

Remove tape and position header WRB Apply self-adhered flashing to seal header WRB

Installation of Sliding Glass Doors

1. **READ INSTRUCTIONS.**
   Read and follow carefully the instructions that come with the door. Never assume what you do not know.

2. **SEAL THRESHOLD.**
   Use neoprene or similar sealing compound to seal the threshold.

3. **INSTALL FLASHING.**
   The jamb flashing with sealer is installed using a method similar to the jamb flashing on windows.

4. **PLACE DOOR.**
   Place the door in position.

5. **CENTER TOP.**
   Center the top of the door in rough opening.

6. **NAIL TOP CORNERS.**
   Nail each corner of the top of the door through nail flange. (Do not set nails.)

7. **ADJUST DOOR.**
   Adjust the door frame so that the space between the door frame and wall trimmers is equal. Check for plumb with a level and adjust if necessary.

8. **COMPLETE NAILING.**
   Close the door and latch it. Then nail off the sides using four 8d nails on each side. Do not nail top of door.

9. **ADJUST DOOR.**
   Adjust the slider part of the door if necessary. Usually there is an adjustment screw at the bottom of the door. Tighten this screw to close a gap between the door and the jamb at the top of the door, or loosen to increase the gap.

10. **TIGHTEN SCREWS.**
    If screws come with the door, shim and tighten screws in the sides and bottom. Use pre-drilled holes.

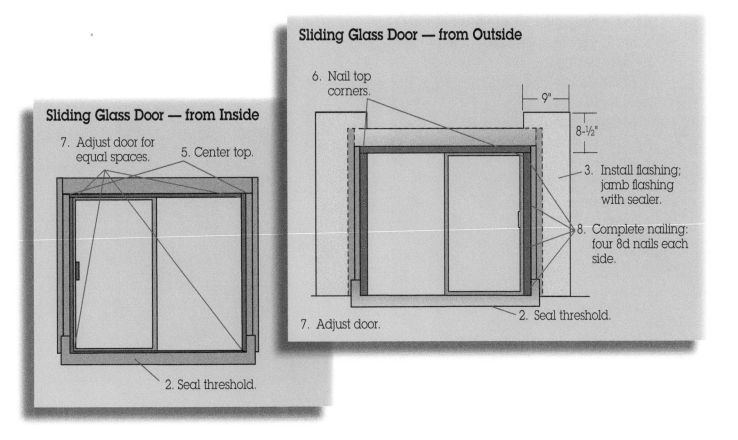

125

Installation of Stairs

The three main dimensions in stair building are for risers, treads, and headroom. The riser height and the tread width are usually given on the plans. You can generally use the tread width given on the plans. The riser height, however, is often not accurate enough to use.

Stair Installation Steps

1. MEASURE HEIGHT.
   Measure the height of the stairwell from finish floor to finish floor.

2. FIND RISER HEIGHT.
   Divide the height of the stairwell by the number of risers shown on the plan to determine the riser height. Be careful to consider the finish floor heights, which may differ top and bottom.

3. FIND TREAD WIDTH.
   Check plans for tread width.

4. CHECK HEADROOM.
   Chalk a line from edge of nosing at top of stairs to edge of nosing at bottom of stairs. (See "Checking Stair Headroom" later in this chapter.) Check for minimum clearance of 6'-8" to finish straight up from line to bottom of headroom.

5. MARK AND CUT STRINGERS.
   (See "Marking Stair Stringers" later in this chapter.)

6. CUT.
   Cut stringer spacers, treads, and risers.

7. NAIL STRINGER SPACER.
   Nail stringer spacer to stringer. Spacer leaves clearance for applying wall finish.

8. SET STRINGERS.
   a. At top deck, measure down riser height plus tread thickness and mark for top of stringer.
   b. Set stringers to mark.
   c. Check stringers for level by placing a tread on top and bottom and checking level, side to side, and front to back.
   d. Adjust stringers for level.
   e. Nail stringers.

9. NAIL RISERS.

10. GLUE AND NAIL TREADS.

Finding Riser and Tread Dimensions

If the riser and tread dimensions are not given on the plans, then you need to calculate them. To do this you should consider the following points:

- You want the steps to feel comfortable.
  - When walking up steps, a person's mind determines the height of the riser based on the first step. Make sure all risers and treads are equal, so the stairs will not cause people to fall.
  - The lower the riser, the longer the tread needs to be to feel comfortable.
- Common dimensions for riser and tread are 7" rise and 10-½" tread.
- Use the following three rules to check to see if your stair dimensions are in the comfortable range.
  - Rule 1: Two risers and one tread added should equal 24" to 25".
  - Rule 2: One riser and one tread added should equal 17" to 18".
  - Rule 3: Multiply one riser by one tread and the result should equal 71" to 75".

Important Stair Code Regulations

The following guidelines for stairs are according to the 2018 *International Residential Code* (IRC) and 2018 *International Building Code* (IBC)

Width –   36" minimum – with occupant load of 49 or less (measured in clear, to finish).

44" minimum – with occupant load of 50 or more (measured in clear, to finish).

Rise –   4" minimum for the IBC.

7-¾" maximum for IRC – 7" for IBC, with residential 7-¾".

Tread –   10" minimum for IRC – 11" for IBC, with residential 10".

Riser height and tread length variance –   ⅜" maximum variation between the treads within any flight of stairs.

Headroom – 6'-8" minimum, measured vertically from a line created by connecting the nosing of the stair treads to the soffit above.

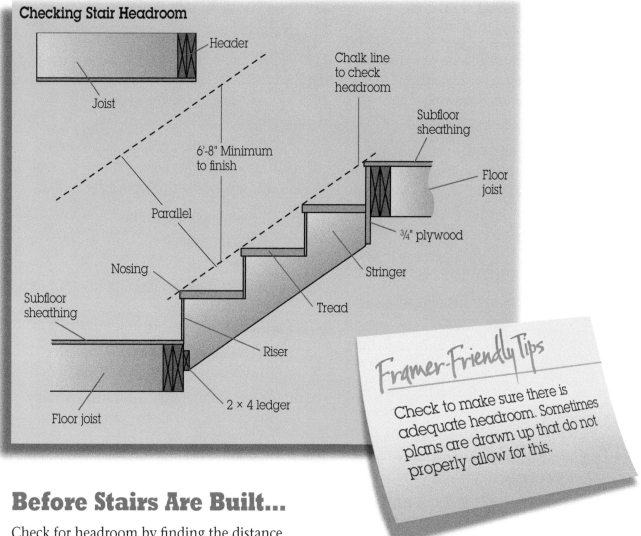

Checking Stair Headroom

Header

Joist

Chalk line to check headroom

Subfloor sheathing

Floor joist

6'-8" Minimum to finish

Parallel

¾" plywood

Nosing

Stringer

Subfloor sheathing

Tread

Riser

Floor joist

2 × 4 ledger

Framer-Friendly Tips

Check to make sure there is adequate headroom. Sometimes plans are drawn up that do not properly allow for this.

Before Stairs Are Built...

Check for headroom by finding the distance vertically between two lines that represent the distance between the stair treads and any obstruction in the headroom. One line is a straight line that connects the nosing on the stairs. The second line is one that runs parallel with the first line but 6'-8" in a vertical direction above the first line.

To create the stair nosing line, first measure up the height of one riser and back from the riser the distance of the nosing and make a mark. From that mark, measure parallel to the subfloor a distance equal to the combined width of the number of treads. Then measure perpendicular to the subfloor

the combined height of the number of risers and make a second mark. Chalk a line between these two marks. From this line, make the second line that is parallel and yet a minimum of 6'-8" vertically. This is your headroom and if anything protrudes into this space you do not have minimum headroom. Remember this distance is to the finish, and so if you are putting carpet or drywall on then you need to allow for their thickness.

Marking Stair Stringers

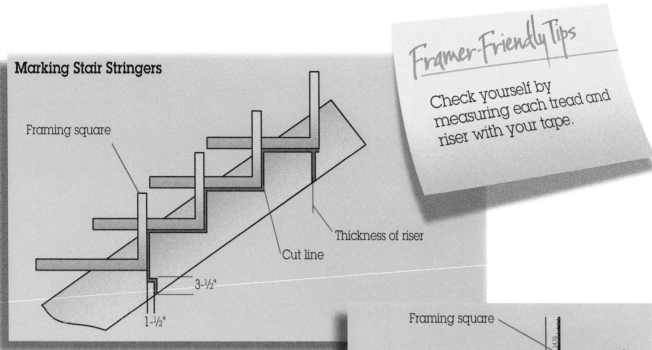

Marking Stair Stringers

Framing square

Thickness of riser

Cut line

3-½"

1-½"

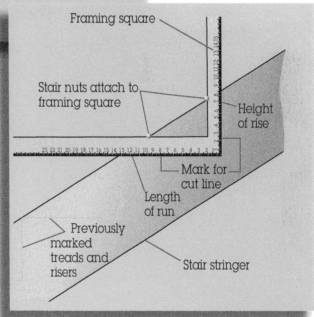

Framer-Friendly Tips

Check yourself by measuring each tread and riser with your tape.

Marking Stair Stringers

1. Measure treads and risers using framing square.

2. Subtract thickness of riser from top.

3. Notch bottom for ledger or top plate. Notches differ. (See illustrations in this and following section.)

4. Care must be taken when marking the top and bottom steps. The thickness of the stair tread and the type of finish flooring on both the tread and the floor must be considered so that all the risers will be the same.

5. For the top tread, be sure to figure in the riser so that the treads and nosings are all equal.

Framing square

Stair nuts attach to framing square

Height of rise

Mark for cut line

Length of run

Previously marked treads and risers

Stair stringer

Framer-Friendly Tips

One of the most common mistakes is not adjusting top and bottom riser height for floor coverings.

Standard Stair
(to be carpeted)

Adapt these guidelines for use when plans do
not give details.

2 × 12 Stair tread:

1. Router nosing with ⅜" round router bit.

2. Glue and nail three 16d nails on each stringer
   for single residence use.

3. Glue and nail four 16d nails on each stringer
   for multi-residence use.

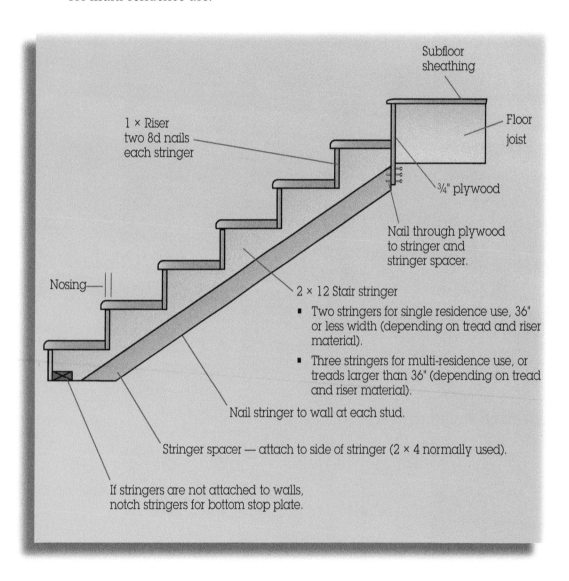

Subfloor
sheathing

Floor
joist

1 × Riser
two 8d nails
each stringer

¾" plywood

Nail through plywood
to stringer and
stringer spacer.

Nosing

2 × 12 Stair stringer

- Two stringers for single residence use, 36"
  or less width (depending on tread and riser
  material).

- Three stringers for multi-residence use, or
  treads larger than 36" (depending on tread
  and riser material).

Nail stringer to wall at each stud.

Stringer spacer — attach to side of stringer (2 × 4 normally used).

If stringers are not attached to walls,
notch stringers for bottom stop plate.

Recap of Key Stair Guidelines

The following are important items to go over to make sure you end up with a good set of stairs:

- Check the code maximum and minimum widths, depths, and heights.

- Remember to review the floor finish on the top, the bottom, and any midway decks for different thickness in the finish floor material. For example, if there is going to be lightweight concrete on the floor sheathing, or if a carpet stair ends on a concrete slab, then the last tread height would have to be adjusted. It is important to stay within the ⅜" height variance (specified in the codes) between all the risers.

- If you have a midway deck in the stairs, make sure you check the height. Figure the height and measure from the top or bottom of the stairs, and then check by figuring the height and measuring from the opposite of top or bottom. If you figured right, your marks should align.

- It is not uncommon for a set of plans to be drawn up with the stair headroom less than the 6'-8" that the code requires. To check the headroom before you frame the stairs, you need to find the point that is plumb, down from the lowest point above the stairs, and then measure to the line in a plane with the nosing of your stair treads.

- Since the stairs are not built yet, the hardest part is finding that nosing plane. You can either work off the plans, if framing has not started, or work with the framing if the frame is ready for the stairs. To find this plumb point on the nosing plane, start from the first nosing, count the number of risers, and multiply that number by the riser height; then add the partial riser.

- To get the partial riser height, you just multiply the partial tread length by the riser percent, which is the riser height, divided by the tread length. Once you have found this length, you can measure either up or down, depending on which direction you used, to see if you have enough headroom.

Curved Stairs

Curved stairs are not as difficult as they seem the first time you think about doing them. They do, however, take some planning and careful work. There is no one way that curved stairs need to be built, as long as they are strong enough to bear the traffic. The method that follows is commonly used.

First of all, unlike straight stairs, we will not use stringers. Instead, each tread will be supported independently, by either a wall or a header. The header method allows for space to be usable under the stairs. The system outlined here uses a header to create what are called *tread walls*.

8 Steps for Building Curved Stairs

1. Find Riser Height

To get started, you first need to find your riser height. Quite often it is given on the plans, in which case you want to check it to make sure it works with the actual floor heights. If the height is not given on the plans, consider the following points when figuring the riser height.

- As with straight stairs, you want the steps to feel comfortable, so remember:
  - Make sure all risers are equal, so the stairs will not cause people to trip and fall.

Drawing the circumference lines

— The lower the riser, the longer the tread needs to be to feel comfortable.

• Common dimensions for riser and tread on straight stairs are 7" for the riser, and 10-½" for the tread. The IRC requires a maximum rise of 7-¾" while the IBC calls for a maximum rise of 7" and a minimum rise of 4".

Because the treads are different for a curved stairway, some new concepts are required. The treads are called winder treads and are defined as having nonparallel edges. The walkline is defined as the line up the treads 12" from where the winder treads are narrower. Winder treads in the IRC have a minimum of 10" at the walkline and a minimum of 6" anywhere else. Winder treads in the IBC have a minimum of 11" at the walkline and 10" anywhere else. Both IRC and IBC required a minimum of ⅜" variation between the greatest and smallest tread widths at the walkline.

2. Mark the Circumference Lines

With the rise figured out and the number of treads known, you can start marking your circumference lines. The best way to start is by making the stair footprint on the floor in the position where the stairs are going to be built. If the plans show a radius dimension and location, then you can use the plans to locate the radius center point. To make your circumference lines (which represent your stairs and the walls on the sides of the stairs), set a nail partway at the located radius center point. Then hook your tape to the nail and mark your circumference lines by swinging your tape around the nail and holding your pencil on the required dimension. (See photo.) Most tape measures have a slot in the hooking end for a nail head. (See illustration later in this section.)

If the radius or the radius center point is not given, you will need to find it. You can vary the radius length, but make sure you can maintain the following three requirements for winder treads:

• IRC minimum 10" at walkline and 6" anywhere else; IBC minimum 10" at walkline and 10" anywhere else

• A minimum stair width of 36" in the clear to finish.

• In non-residential buildings, the smaller radius should not be less than twice the width of the stairway or required capacity of the stairway.

Framer-Friendly Tips

It's a challenge to hold your pencil on the tape while marking. It's easy to draw a second line to check your accuracy.

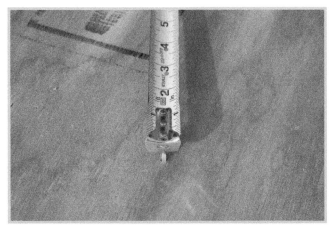

End of tape hooked to a partially set nail

The first thing you need to do to find your radius is to establish two points on the circumference opposite each other. They can be any two points. Look on the plans for points that are already established. If there are no established points, then select points that fit with the location of the stairs. Once you have established two points, it is merely a matter of bisecting the line between these points, finding the radius origin, and drawing your circumference lines from the radius origin. (See the "Bisecting a Line to Establish the Radius Origin" illustration.)

3. Marking the Tread on the Footprint

Now that you have your circumference lines, you need your tread lines. Since you have figured your riser height, you know the number of treads that you will have. Knowing the number of treads, you can find the exact tread point along your stair circumference. To do this, divide the stair circumference in half, and then divide those halves in half again and again until you are down to single treads. (See "Divide Circumference for Treads" illustration.)

If your stair has an uneven number of treads, then you have to subtract one tread before you begin dividing into halves. To subtract one tread, you first have to know the width. The width will be equal to the total stair circumference length divided by the number of treads.

It is difficult to measure the stair circumference. If your tread is not exact, use the information gained from the first marking and remark the division point.

Once you have all your division points for the treads, then chalk lines from the radius center point through the division points to the longest circumference line, and those lines will make your tread footprint. (See "Tread Footprint" illustration.)

4. Cut Bottom Plate

The bottom plate of the tread walls will not be parallel to the top plate, as it would be in a straight stair. The bottom plate will follow the circumference and serve as the bottom plate for all the tread walls. A good way to make the bottom plate is to use two pieces of ¾" plywood. If the radius is not too small, you can cut the plywood with a circular saw. To mark on the plywood, set a nail anywhere and mark the plywood with a pencil and a tape measure. Use the dimensions from the stair footprint to get the radius length.

5. Nail Bottom Plate in Place

To build the stairs, start by nailing the bottom plates in place. (See "Bottom Plate Nailed in Place" illustration.)

6. Build the Tread Walls

The walls supporting the treads will be built as header walls. Built this way, they will provide the riser and allow space for storage below the stairs. The wall will consist of a 2 × 12 single header that will serve as the riser, a top plate, a double plate, trimmers for under the 2 × 12 header, and king studs next to the trimmer. A ledger to support the tread below will be nailed onto the header. (See "Section of Tread Wall from End" illustration.)

Each tread wall should be higher than the one below it by the riser height. The height of the first step will have to be figured separately to equal one riser height, adjusted for any difference in floor covering. The top step might also have to be adjusted for a difference in floor height.

7. Install the Tread Walls

Nail the tread walls in place using the footprint lines. The bottom of the studs will be toenailed into the bottom plate already in place. (See "Tread Walls Nailed in Place" illustration.)

8. Cut and Nail Treads

The treads should all be the same. They will be nailed onto the top of the tread walls and the ledgers. An equal nosing should be maintained the full length of the tread. Make sure the walls stay plumb both ways. Glue each tread to prevent squeaks. (See "Treads Halfway Up Stairs" illustration.)

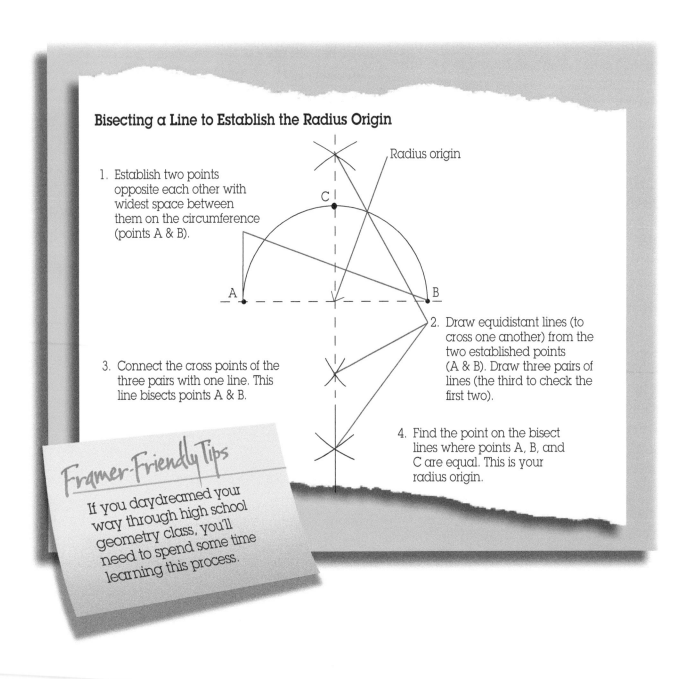

Bisecting a Line to Establish the Radius Origin

1. Establish two points opposite each other with widest space between them on the circumference (points A & B).

2. Draw equidistant lines (to cross one another) from the two established points (A & B). Draw three pairs of lines (the third to check the first two).

3. Connect the cross points of the three pairs with one line. This line bisects points A & B.

4. Find the point on the bisect lines where points A, B, and C are equal. This is your radius origin.

Radius origin

Framer-Friendly Tips

If you daydreamed your way through high school geometry class, you'll need to spend some time learning this process.

Divide Circumference for Treads

1. Draw lines from points A & B to the point where the bisecting line intersects the circumference (at point C) creating half lines.

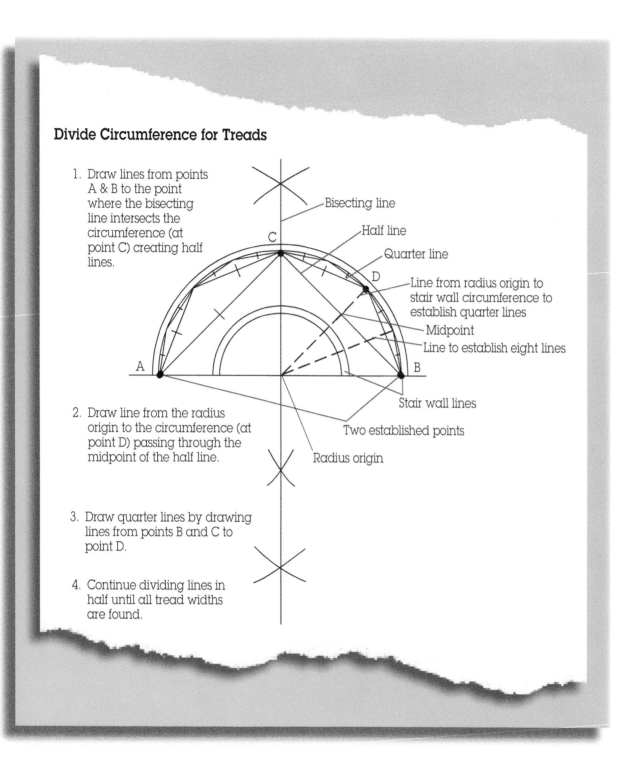

Bisecting line

Half line

Quarter line

Line from radius origin to stair wall circumference to establish quarter lines

Midpoint

Line to establish eight lines

Stair wall lines

Two established points

Radius origin

2. Draw line from the radius origin to the circumference (at point D) passing through the midpoint of the half line.

3. Draw quarter lines by drawing lines from points B and C to point D.

4. Continue dividing lines in half until all tread widths are found.

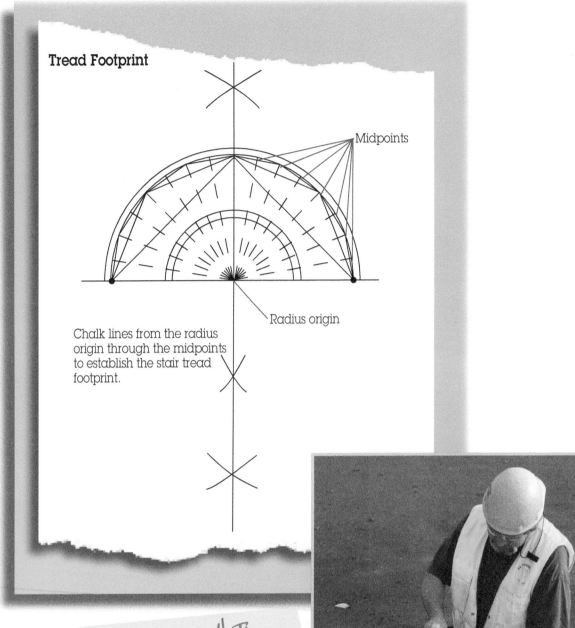

Tread Footprint

Midpoints

Radius origin

Chalk lines from the radius
origin through the midpoints
to establish the stair tread
footprint.

Framer-Friendly Tips

On smaller circumferences,
you'll occasionally want
to back up your saw as you
cut to follow the line.

Cutting curved bottom plate

Bottom Plate Nailed in Place

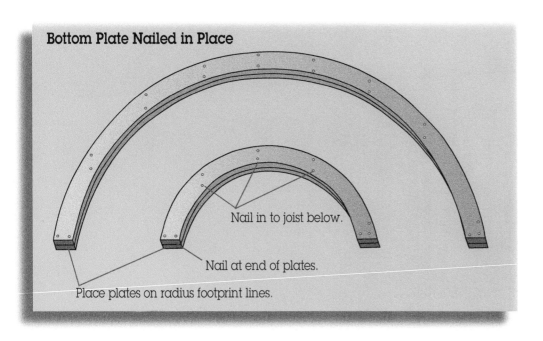

Nail in to joist below.

Nail at end of plates.

Place plates on radius footprint lines.

Section of Tread Wall from End

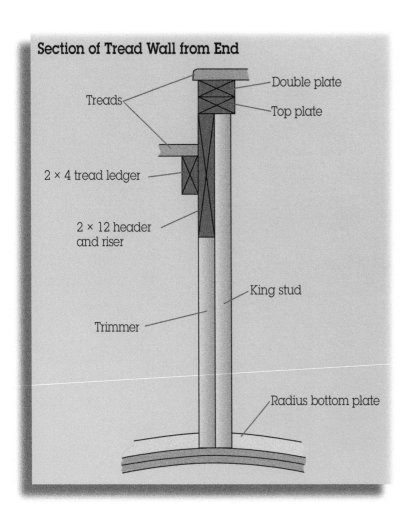

Treads

Double plate

Top plate

2 × 4 tread ledger

2 × 12 header and riser

King stud

Trimmer

Radius bottom plate

137

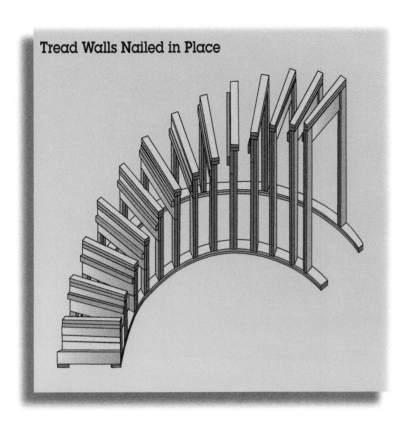

Tread Walls Nailed in Place

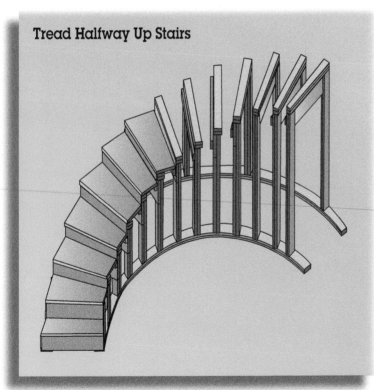

Tread Halfway Up Stairs

138

Chapter Seven
LAYOUT

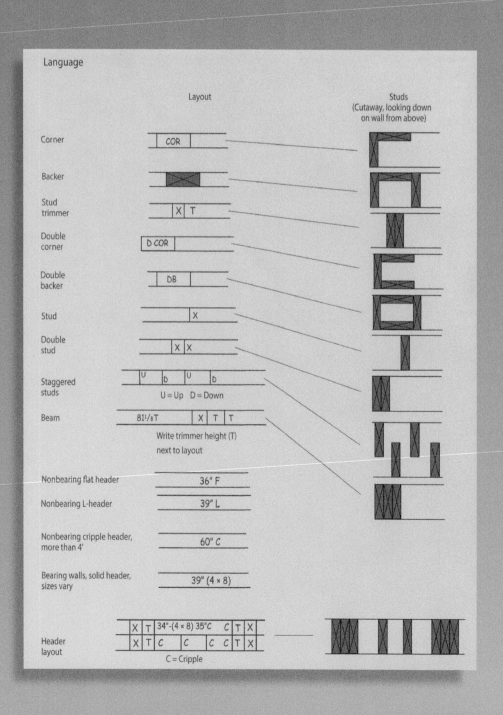

Language

	Layout	Studs (Cutaway, looking down on wall from above)
Corner	COR	
Backer		
Stud trimmer	X · T	
Double corner	D COR	
Double backer	DB	
Stud	X	
Double stud	X · X	
Staggered studs	U · D · U · D (U = Up D = Down)	
Beam	81¹/₈T · X · T · T Write trimmer height (T) next to layout	
Nonbearing flat header	36" F	
Nonbearing L-header	39" L	
Nonbearing cripple header, more than 4'	60" C	
Bearing walls, solid header, sizes vary	39" (4 × 8)	
Header layout	X · T · 34"-(4 × 8) 35"C · C · T · X / X · T · C · C · C · C · T · X (C = Cripple)	

Contents

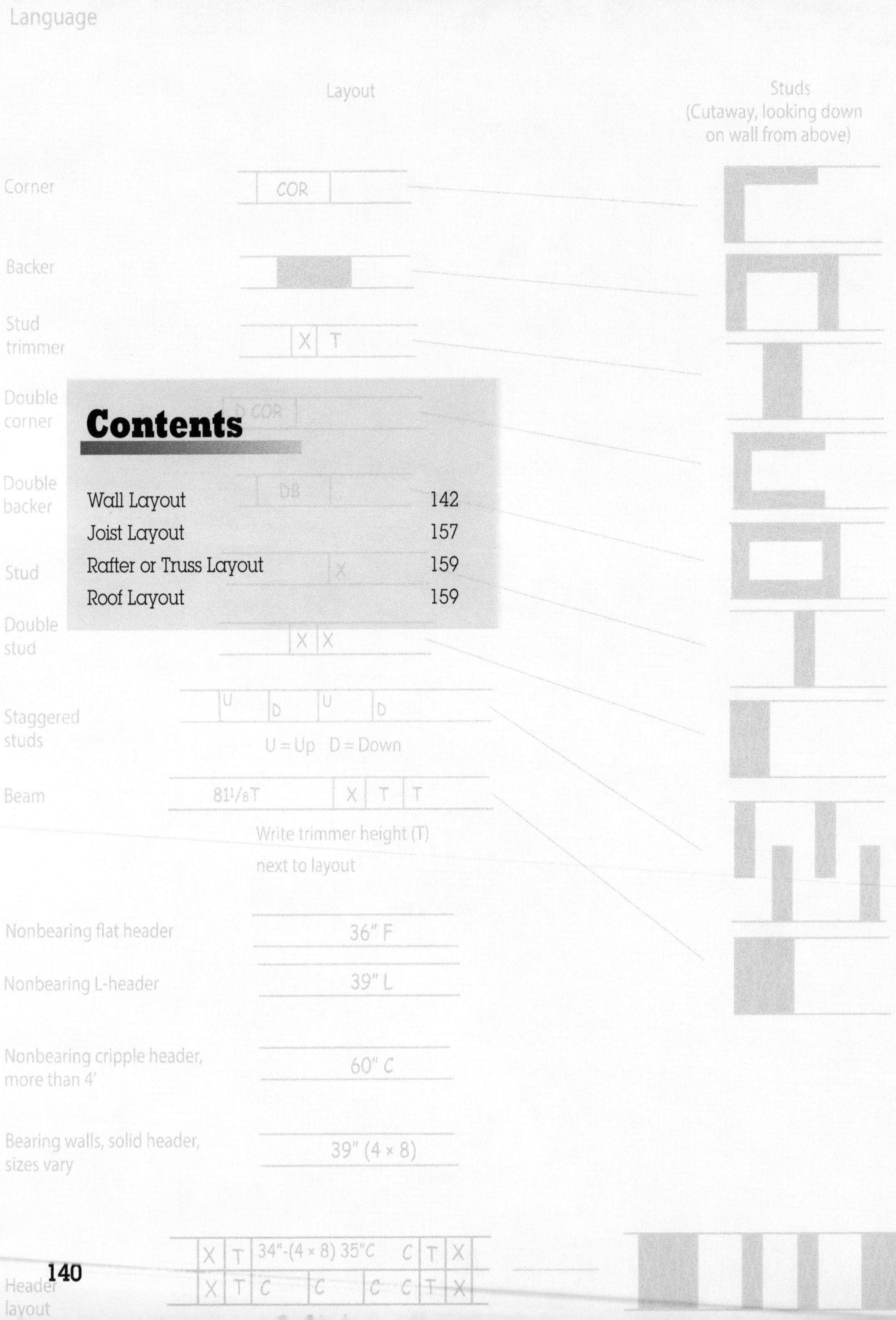

Language
Layout
Studs (Cutaway, looking down on wall from above)
Corner
COR
Backer
Stud trimmer
X T
Double corner
D COR
Double backer
DB
Stud
X
Double stud
X X
Staggered studs
U D U D
U = Up D = Down
Beam
81¹/₈T
X T T
Write trimmer height (T) next to layout
Nonbearing flat header
36″ F
Nonbearing L-header
39″ L
Nonbearing cripple header, more than 4′
60″ C
Bearing walls, solid header, sizes vary
39″ (4 × 8)
X T 34″-(4 × 8) 35″C C T X
X T C C C C T X
Header layout

Chapter Seven
LAYOUT

Layout is the written language of the framer. If the lead framer on the job "writes" clearly, then the framers reading the layout will be able to understand and properly perform the work. It's important to include enough information in the layout so that there aren't any questions. Layout language has been developed by framers over the years, and there are some variations. The version described in this book is quite typical. Feel free to make any changes that reflect practices in your area. If you need to explain something about the layout that isn't shown in this chapter, either write it out on the plates in plain language, or explain it to the person who will do the framing.

Layout for framing requires bringing together the desires of the owner, the written instructions of the architect and engineer, instructions from the builder and/or superintendent, and materials from the supplier—then writing these instructions on job-site lumber in a legible manner so that the framers can build the walls, floors, and roofs without continuous interpretation. This chapter describes this process and explains the written words and symbols the lead framer uses.

The approach you use will depend on the size of the job, the area of the country you are working in and, most of all, the style of the person who taught you framing layout. This chapter presents a common style of layout, with some variations. Any style you use is good, as long as the framers can read and understand it, and you have provided all the information they need to frame the building completely.

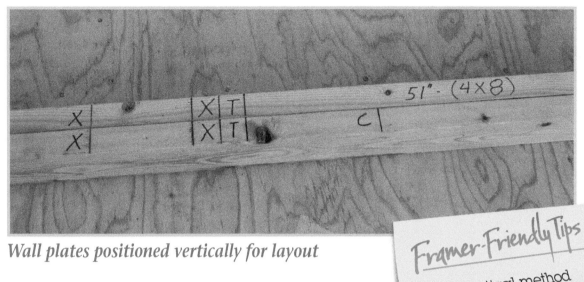

Wall plates positioned vertically for layout

Framer-Friendly Tips

The vertical method makes layout easier.

Wall Layout

On many jobs, the basic skeleton of the walls is built, and then the blocking, hold-downs, and miscellaneous framing are filled in as a later operation. There are some disadvantages to installing miscellaneous framing after the basic framing. For example, you may have to notch around wires and pipes. You may even have to come back and set up a separate operation after you have already left the job site. This chapter describes a system that includes everything possible in the layout, so the walls can be framed, complete, all at one time. To do this takes organization and pre-planning, which includes gathering all the information you need before you do the layout.

The positioning of the top plate and bottom plate for layout detailing is a variable that depends on personal preference and the type of operations. The plate can be positioned vertically so that the 1-½" width is on top, which makes it easy for marking on the plate. The plate can also be laid flat (horizontally) on top of the chalk lines so that the plates are in the same position as when the walls are standing. This system makes it easy to keep the walls in the proper position, particularly when you have angled walls. A third option (for some exterior walls only) is to position the bottom plate where it will be once the wall is standing, then tack the top plate to it, hanging over the side. This system works well if you want to attach the bottom plate to the floor and then stick-frame the wall.

The layout language varies, but all layout styles are similar. The chart on page 139 shows the basic layout language. Page 144 and 147 shows additional language. Although the parts of the walls are typically the same in different areas of the country, quite often they are referred to by different names. For example, a *backer* is also known as a *channel* or *partition*. Even the term "layout" can have different meanings. Sometimes layout is understood to be the total process of chalking the lines for the wall locations (snapping), cutting the plates, and writing the layout language on the plates (detailing). It is not important what terms are used, as long as there is clear communication.

Wall layout is the process of taking the information given on the plans and writing enough instructions, in layout language, on the top and bottom plates so the framer can build the wall without asking any questions.

Following is some general information that must be considered before starting. Unless otherwise noted, all layout discussions will assume 2 × 4 studs at 16" O.C. (on center).

Where possible, we want joists, studs, and rafters to rest directly over each other.

Before layout is started, establish reference points in the building for measuring both directions of layout and use those points for joist, stud, and rafter/truss layout throughout the building.

Check the building plans for a special joist plan or rafter/truss plans indicating layout.

Select a reference point which allows you to lay out in as long and straight a line as possible, and which ensures that a maximum number of rafters/trusses are directly supported by studs.

Wall Layout Steps

1. Spread the top plate and bottom plate together in chalk lines. If a plate is not long enough, cut the top plate to break on the middle of the stud and four feet away from walls running into it.

2. Place plates in position with chalk lines.

3. Lay out for backers from chalk lines.

4. Lay out stud trimmers and cripples for windows and doors.

5. Lay out studs.

Chalking Lines

"Chalking lines" is the process of marking on the subfloor where the walls are to be placed. Red chalk makes a permanent line and is easily seen. Blue chalk can be erased and is good to use if the lines might have to be changed. Using different colors allows you to distinguish between old and new lines.

Before chalking, when possible, check foundation and floor for square. Walls must be square, plumb, and level. If necessary, adjust your chalk lines accordingly.

Measurements for chalk lines are derived from wall dimensions as given in the plans. If the plans show finished walls, be sure to subtract the appropriate amount to get your framing measurements.

Miscellaneous Wall Framing Layout

Each building has unique characteristics that require special attention. Hold-downs, shear walls, blocking, backing, special stud heights, and posts are some of the more common miscellaneous framing items. The framing language for these framing items is not well-defined because the operations frequently change and the same ones are not always used. The Miscellaneous Layout Language Chart on the next page gives you an idea of how these framing tasks can be communicated.

Hold-Downs

Hold-downs are probably the most difficult to mark correctly. They vary from location to location and require different studs or posts for connecting.

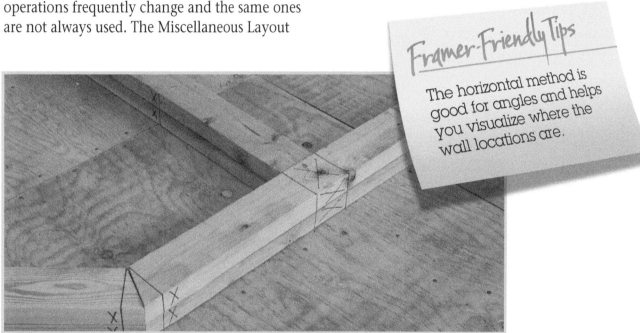

Framer-Friendly Tips

The horizontal method is good for angles and helps you visualize where the wall locations are.

Wall plates positioned horizontally for layout

Framer-Friendly Tips

The hanging plate method is good if the bottom plate is attached to the floor.

Top plate tacked to bottom plate hanging over edge of concrete for layout

Miscellaneous Layout Language Chart

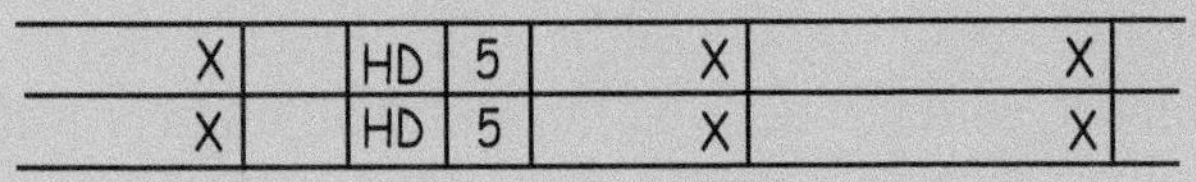

X	HD	5	X		X
X	HD	5	X		X

Hold-down 5

X	6P4	X	X
X	6P4	X	X

6" O.C. plate nailing

P = Plywood sheathing

4" edge nailing

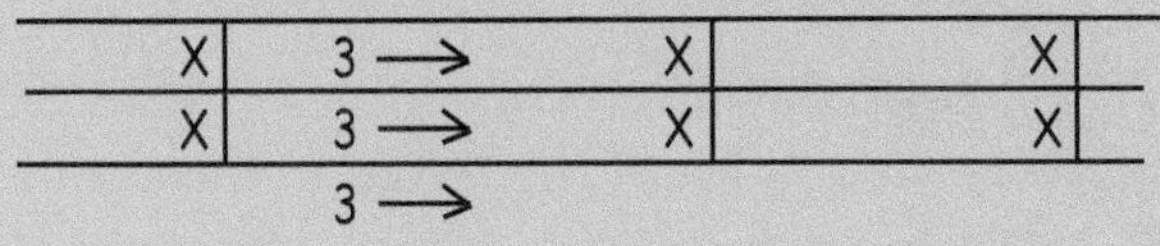

X	3 →	X	X
X	3 →	X	X
	3 →		

Location mark

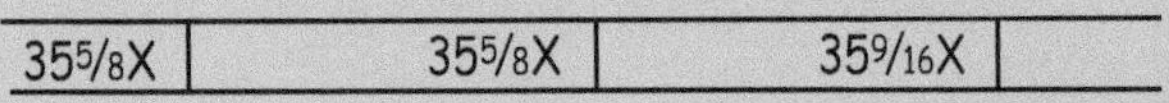

$35^5/_8$X	$35^5/_8$X	$35^9/_{16}$X

Special length studs

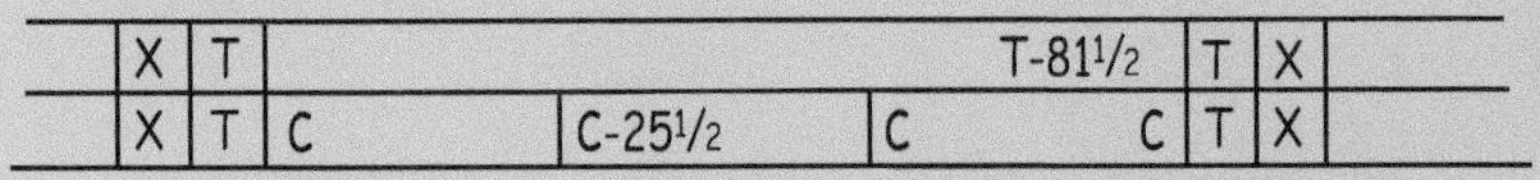

X	T			T-81½	T	X
X	T	C	C-25½	C	C	T	X

Trimmer heights
Cripple heights

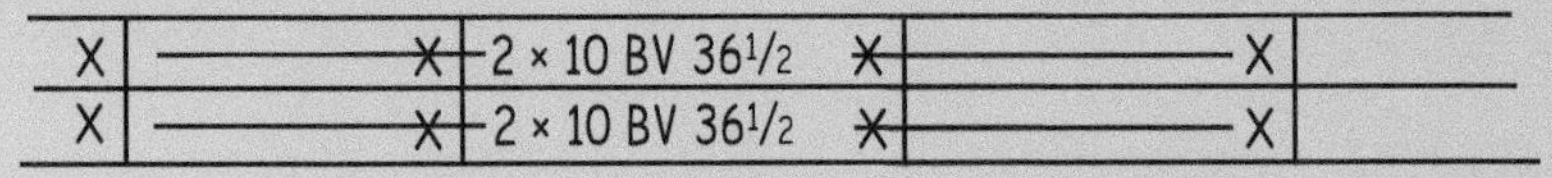

Blocks

2 × 10

B = Block

V = Vertical

36-½" = Center height from floor

——— = extent of blocking

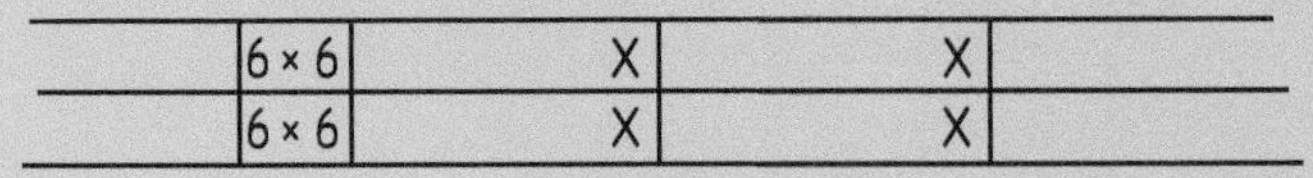

X	——×—2 × 10 BV 36½ ×————	X
X	——×—2 × 10 BV 36½ ×————	X

6 × 6	X	X	
6 × 6	X	X	

Posts

X		X	X
X		X	X

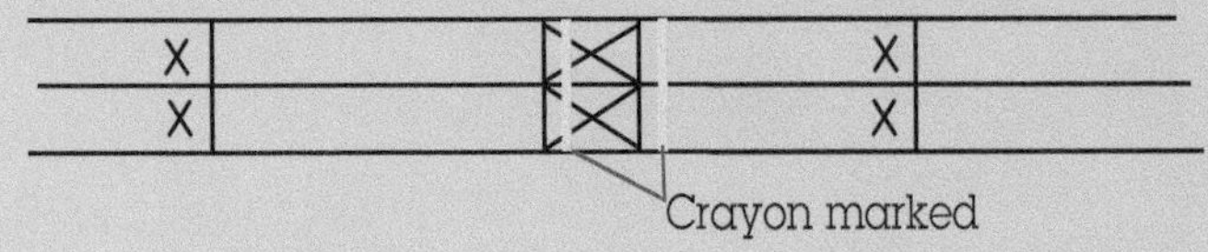

Offset backer

Crayon marked

There are also different types of hold-downs, and each manufacturer has its own identification system. There are, however, four basic styles of hold-downs that you need to show in your wall layout.

1. Basic hold-down: bolts, nails, or screws to the hold-down post or studs. This type is typically attached to an anchor bolt in the foundation or bolted to an all-thread rod that is connected to a hold-down in the wall below. (See "Hold-Downs with Floor Between" illustration.)

2. Hold-down already embedded in the concrete, which needs only to be attached to the wall. (See "Hold-Down in Concrete" illustration.)

3. Strap used to connect the top of one wall to the bottom of the wall above. (See "Strap Wall to Wall" illustration.)

4. Hold-down that is continuous between all floors from the foundation to the top floor. (See "Hold-Down, Continuous" illustration.)

The difficulty in laying out for hold-downs is knowing what to write on the plates so that the requirement will be easily understood. The best thing to do is to explain to all framers at the beginning of each job what symbols you are using to indicate hold-downs. Use the same language when possible at different jobs. The most common symbol for hold-downs is HD, followed by the number representing the size of the hold-down—for example, HD2 or HD5.

When you are laying out for hold-downs, it's important to get the layout in the right location. Since the purpose of a hold-down is to connect the building to the foundation, the hold-downs must

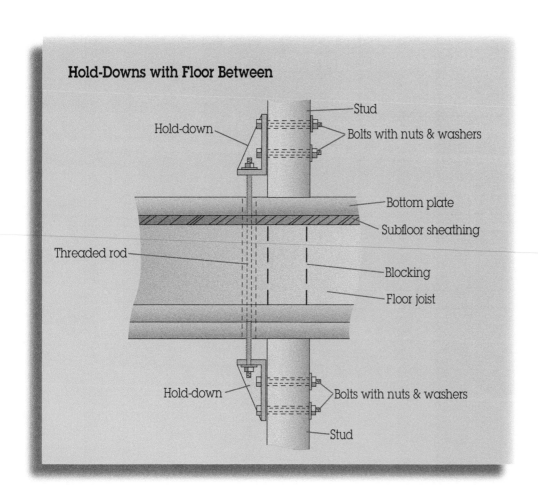

Hold-Downs with Floor Between

Stud

Hold-down

Bolts with nuts & washers

Bottom plate

Subfloor sheathing

Threaded rod

Blocking

Floor joist

Hold-down

Bolts with nuts & washers

Stud

Hold-Down in Concrete

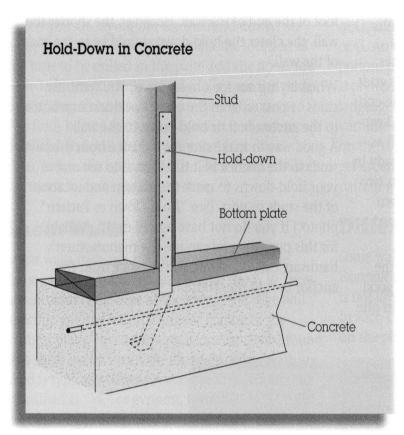

Stud

Hold-down

Bottom plate

Concrete

Strap Wall to Wall

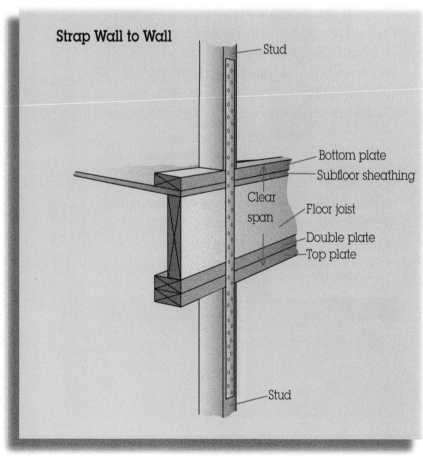

Stud

Bottom plate

Subfloor sheathing

Clear span

Floor joist

Double plate

Top plate

Stud

Backer Move

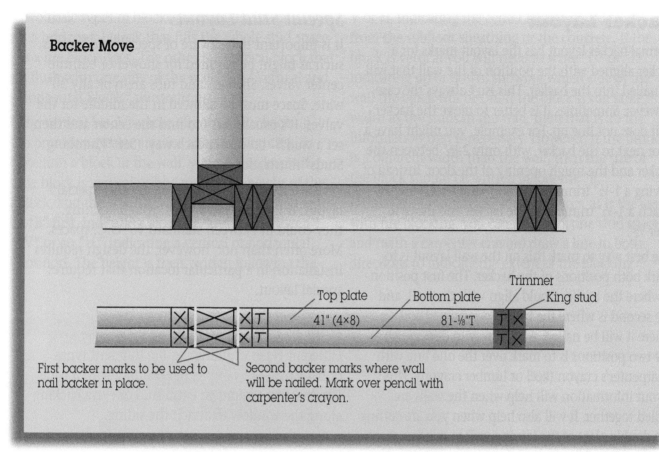

Top plate

Bottom plate

Trimmer

King stud

41" (4×8)

81-⅛"T

First backer marks to be used to nail backer in place.

Second backer marks where wall will be nailed. Mark over pencil with carpenter's crayon.

Plumbing Studs

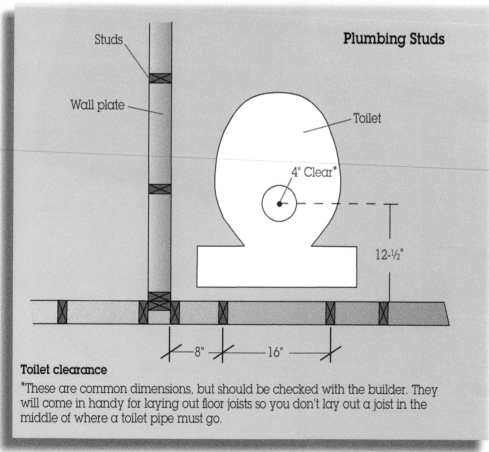

Studs

Wall plate

Toilet

4" Clear*

12-½"

8"

16"

Toilet clearance

*These are common dimensions, but should be checked with the builder. They will come in handy for laying out floor joists so you don't lay out a joist in the middle of where a toilet pipe must go.

Structural Support

Often a beam or girder truss, or some other structural member, requires structural support all the way to the foundation, but that support is not shown on the plans. Check and lay out for upper floor structure support when laying out walls.

Location Marks

Another mark you might need to make on your walls is for location. If you have to move the plates to make room to build walls, then you need to mark the location for all the walls to make sure you know where they go when you are ready to build them. It is best to use a crayon and mark a number and an arrow on each plate. The top and bottom plates will be marked with the same number. The arrows will point in the same direction as all the other walls that run in that direction. In addition to marking the plates, mark the number and the arrow on the floor next to the plates. (See "Plate location marks" photo.)

If you are laying out walls on a concrete slab, then you will have to contend with plumbing and electrical pipes. Before you start your layout, notch the bottom plate to fit.

Layout Methods

Use the Correct Order

When you perform the layout, follow a prioritized order. For example, trimmer and king studs for doors and windows take priority over studs. That is because if a stud falls on the location of a trimmer or king stud, then the stud is eliminated. Using a certain order for layout also helps you keep track of where you were if you are pulled off layout and have to come back later to pick up where you left off. The order should be doors and windows first, then bearing posts, backers and corners, then hold-downs—followed by special studs like medicine cabinet studs, then regular studs, and finally miscellaneous framing, such as blocks.

Align Framing Members

It is good practice when laying out studs to align the roof trusses, floor joists, and studs. This is not entirely possible in most cases because the studs are typically 16" O.C., while the roof trusses or rafters are 24" O.C. However, you will at least line up every third truss or rafter. If the studs are 24" O.C., then they need to align with the trusses or rafters. Aligning the studs, joists, and trusses or rafters not

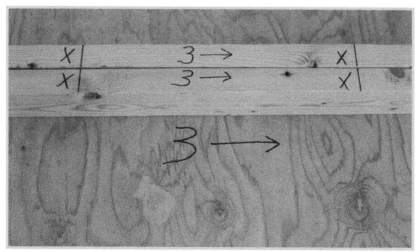

Plate location marks

only makes good sense structurally, but makes it easier for the plumbers and electricians to run their pipes and wire between floors.

Consult the Roof and Floor Drawings

To start laying out your studs, you need to know the layout of your roof and floor systems. If you are using dimensional lumber, then you can most likely start the layout for your roof and floor wherever it is convenient. If, however, you are using I-joists or roof trusses, the layout will probably be defined on a set of shop drawings provided by the supplier. You should receive a set of these shop drawings before you start laying out your studs, so that you can align wherever possible. Once you have decided on a starting point for your layout in each direction, use the same layout throughout the building. Although it is not structurally necessary to align nonbearing interior walls with multiple floors, it is helpful to have the studs aligned for the plumbers and electricians.

A Word of Caution

Some production framing techniques speed up the layout process, but be careful, if you use them, not to sacrifice quality. For example, instead of using the X with a line next to it to indicate a stud, a single line can be used to represent the center of the stud. Be careful with this designation, because the studs need to line up with the middle of the wall sheathing. If you figure that you allow $\frac{1}{8}$" for expansion between the sheets of sheathing, that only allows $1\frac{1}{16}$" for nailing each side. You cannot afford to be off by even a small amount and still get enough stud to nail to. If you use this system, you also have to be sure that your framers are competent and can align the studs properly.

Other Tasks that Can Be Done Along with Layout

Some items can be attended to while you are performing the layout. One is to cut a kerf in the bottom of the bottom plate at door thresholds when they are sitting on concrete. This kerf (about half the thickness of the plate) allows you to cut out your bottom plate after the walls are standing without ruining your saw blades on the concrete. (See "Kerf cut [threshold cut]" photo.)

You can take care of another item while drilling the bottom plate to install over anchor bolts. When the bottom plate is taken off the bolts to do the layout, it can be turned over and accidentally built into the wall upside-down. This problem can be prevented by using a carpenter crayon to mark "UP" on the top

Kerf cut (threshold cut)

of the plate before it is removed from the bolts. Angle walls have the same potential for getting built with the plates backward. You can also mark them as they are being laid out.

Layout Tools

There are some tools that can help with layout, particularly with multi-unit or mass production-type framing. One of these is the channel marker, a template made to assist in laying out corners and backers. Another is a layout stick. The layout stick is 49-½" long and is placed on the plates to act as a jig for marking studs. (See "Channel marker and layout stick" photo.)

Joist Layout

Floor Joist

Joist layout is relatively easy compared with wall layout. It uses the same basic language as walls.

Special layout for joists includes the area under the toilet and shower drain. It is easier to move a joist a couple of inches or even to add a joist than to come back and "header-out" a joist because the plumber had to cut it up to install pipes. On larger buildings, there may be shop drawings for the floor joists that can be used for your layout. The shop drawings should show the locations of openings and how they should be framed. (See "Joist Layout Language" illustration.)

Channel marker and layout stick

Joist Layout Principles

Joist layout is the process of taking the information given on the plans and writing enough instructions on the top of the rim joist or double plate so the joist framer can spread and nail the joists without asking any questions.

Where possible, we want joists, studs, and rafters to set directly over each other.

Before layout is started, establish reference points in the building for measuring both directions of layout and use those points for joist, stud, and rafter/truss layout throughout the building. Check the building plans for special joist plan or rafter/truss plans indicating layout. Select a reference point which allows you to lay out in as long and straight a line as possible, and which ensures that a maximum number of rafters/trusses are directly supported by studs.

Check plans for openings in the floor required for stairs, chimneys, etc.

Check plans for bearing partitions on the floor. Double joists under bearing partitions running parallel.

Check locations of toilets to see if joists must be headed-out for toilet drain pipes.

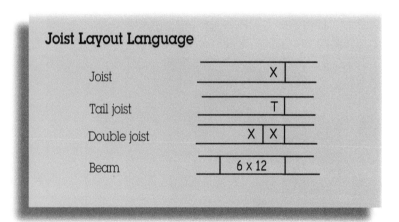

Joist Layout Steps

1. Layout for double joist, trimmer joist, and tail joist.
2. Layout for other joists.

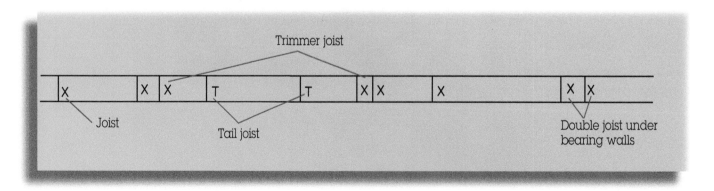

Rafter or Truss Layout

Rafter or truss layout, like floor joist layout, is relatively easy compared with wall layout. Sometimes it is helpful to lay out for rafters or trusses on the top of the double plate so that once the wall is standing, the layout will already be done, and you won't have to do it from a ladder. (See "Rafter layout on walls before the wall is stood up" photo.)

Special layout is often required for ceiling can-lights. Check on necessary clearance to make sure you provide enough room.

Roof Layout

Roof layout is the process of taking the information given on the plans and writing enough instructions on the double plate for the roof framer to spread and nail the rafters or trusses.

Use the same reference points established for floor and wall building for starting layout on the roof.

Roof rafters and trusses are sometimes 24" O.C. as compared with 16" O.C. for floors and walls. In that case, only every third truss or rafter will be over a stud.

Before layout is started, check plans for openings in the roof required by dormers, skylights, chimneys, etc.

For roof trusses, lay out according to truss plan, especially for hip-truss packages.

Roof Layout Language

Rafter or truss	X
Tail rafter or truss	T
Double rafter or truss	X X

Roof Layout Steps

1. Lay out for doubles, trimmers, and tail rafters or trusses.
2. Lay out for other rafters or trusses.

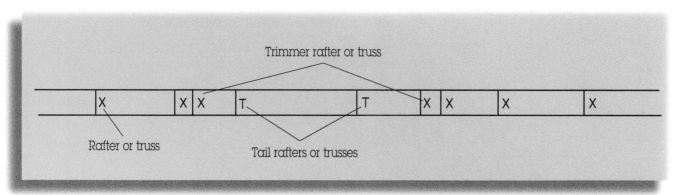

Trimmer rafter or truss

Rafter or truss

Tail rafters or trusses

Rafter layout on walls before the wall is stood up

Framer-Friendly Tips

Be creative with your layout language as needed. Just make sure the framer understands it.

Conclusion

Writing layout is just like writing anything else. If the person reading it understands what you want to say, then you've done a good job. When you are done with the layout, take a look at it to make sure you can read it. If your writing is not showing up clearly, you might try a different brand of carpenter pencil. There are different leads, and some write better than others, depending on the condition of the wood. You can also use indelible marking pens, which are especially good on wet lumber.

Review the layout with the framer who will be reading it before he starts. If a framer doesn't understand your layout, it takes more time for him to try to figure it out than for you to explain it to him. A little extra time spent on layout is usually a good investment. It's not easy taking information from all the different sources that combine in the construction of a building and making it legible for framing, but if the layout is communicated clearly, it will help the framers do their work in an organized and productive manner.

Chapter Eight
ENGINEERED WOOD PRODUCTS

Contents

ENGINEERED WOOD PRODUCTS

Engineered wood products have been around for years, particularly in the form of plywood, glu-lam beams, and metal-plate-connected wood trusses. I-joists are more recent, as are LVL (laminated veneer lumber), PSL (parallel strand lumber), LSL (laminated strand lumber), OSL (oriented strand lumber), and CLT (cross-laminated timber).

It is not the intent of this chapter to explain everything there is to know about engineered wood products, but rather to make you familiar with this category of materials, and give you a sense of what to look out for when you are working with them.

Engineered wood products (EWP) fit into two general categories, **engineered panel products (EPP)** and **engineered lumber products (ELP)**. The first group includes plywood, oriented strand board (OSB), waferboard, and composite and structural particleboard.

The second group includes I-joists, glu-lam beams, metal-plate-connected wood trusses, and structural composite lumber (LVLs, PSLs, and LSLs).

Engineered Panel Products

Engineered panel products are so common that their uses are defined in the building codes. Specific applications vary from job to job, and from manufacturer to manufacturer.

Oriented Strand Board and Waferboard

Most building codes recognize oriented strand board and waferboard for the same uses as plywood, as long as the thicknesses match.

Working with Engineered Panel Products

When working with any engineered panel products, keep the following guidelines in mind:

1. On floors and roofs, run the face grain perpendicular to the supports (except with particleboard, which has no grain). See "Using Engineered Panel Products" illustration.

2. Do not use any piece that does not span at least two supports for floors and roofs.

3. Allow a gap of at least ⅛" on all edges, and a gap of more than ⅛" if the piece will be exposed to a lot of moisture before the siding is installed. Note that this also applies to walls.

4. Follow manufacturers' recommended installation directions.

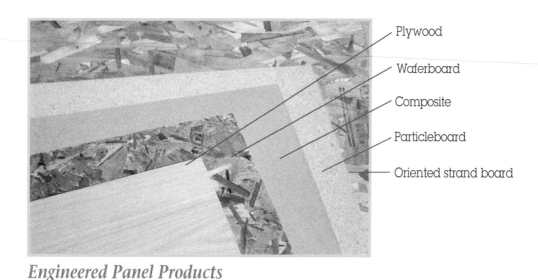

— Plywood

— Waferboard

— Composite

— Particleboard

— Oriented strand board

Engineered Panel Products

Engineered Lumber Products

I-Joists

I-joists were introduced in 1968 by the Trus Joist Corporation. Although use of this product has grown rapidly over the years, there is still no industry standard for its manufacture and installation. And while the Engineered Wood Association (APA) has established a standard for its members, not all manufacturers are members of APA. Because there is no universal standard, it's important to use the installation instructions that come in the I-joist package. The package is generally prepared by a manufacturer's representative working with the architect or designer.

The I-joist package should include installation plans for the building. These plans will be specific to the building you are working on, and will include a material list and accessories. Accessories can include web stiffeners, blocking panels, joist hangers, rim boards, and beams. The plans typically include a sheet of standard details. The following is a list of elements you'll find in most I-joist packages, and some items to consider when installing them:

1. **Minimum bearing** is 1-¾". (See "Solid Blocking and I-Joist Minimum Bearing" illustration.)

2. **Closure** is required at the end of the I-joist by rim-board, rim-joist, or blocking. This closure also serves to transfer vertical and lateral loads, as well as providing for deck attachment and fireblocking, if required. Do not use dimensional lumber, such as 2 × 10, because it is typically 9-¼" instead of 9-½". It shrinks much more than the I-joists and will leave the I-joists supporting the load.

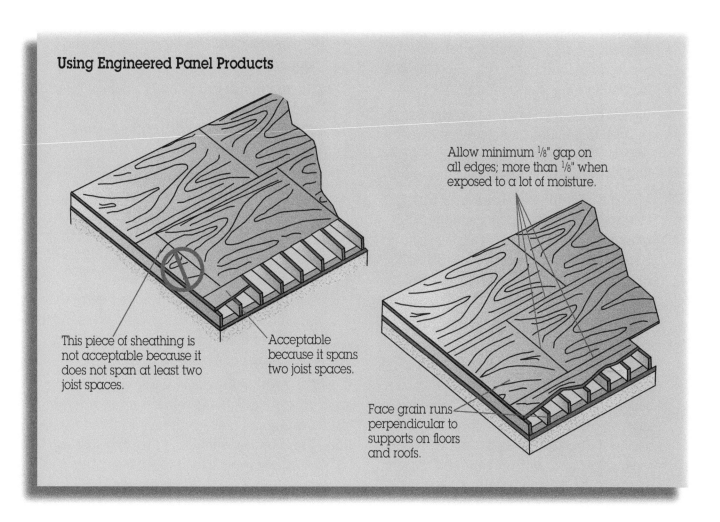

Using Engineered Panel Products

Allow minimum ⅛" gap on all edges; more than ⅛" when exposed to a lot of moisture.

This piece of sheathing is not acceptable because it does not span at least two joist spaces.

Acceptable because it spans two joist spaces.

Face grain runs perpendicular to supports on floors and roofs.

3. **Interior bearing walls** below I-joists require blocking panels or squash blocks when load-bearing walls are above. (See "Interior Bearing Wall Blocking Panel" illustration.)

4. **Rim boards** are required to be a minimum of 1-¼" in thickness.

5. Make sure **squash blocks,** which are used to support point loads (like the load created by a post), are ¹⁄₁₆" taller than the joists, so that they will properly support the load. (See "Squash Blocks" illustration.)

6. **Web stiffeners**, which are sometimes required at bearing and/or point loads, should be at least ⅛" shorter than the web. Install web stiffeners tight against the flange that supports the load. If the load comes from a wall above, install the web stiffener tight against the top of the flange. If the load comes from a wall below, the stiffener should be installed tight against the bottom. (See "Web Stiffener" illustration.)

7. Use **filler blocking** between the webs of adjacent I-joists to provide load sharing between the joists. (See "Filler Blocking and Backer Blocking" illustration.)

8. **Backer blocking** is attached on one side of the web to provide a surface for attachment of items like face-mount hangers. (See "Filler Blocking and Backer Blocking" illustration.)

9. I-joists are permitted to **cantilever** with very specific limitations and additional reinforcement. If the I-joists are supporting a bearing wall, the maximum cantilever distance with additional reinforcement is 2'. Check the plans for specifics on the cantilever.

10. **Top-flange hangers** are most commonly used for I-joists. (See "Top Flange Hanger Tight" illustration.) They come with the I-joist package, but you can also get them from a construction supply store. When installing top flange hangers, make sure that the bottom of the hanger is tight against the backer block or the header. When nailing the hanger into the bottom of the joist, be sure to use the correct length nails. Nails that are too long can go through the bottom flange and force the joist up. (See "Use Right Size Nail" illustration.) When installing hangers on wood plates that rest on steel beams, the hanger should not touch the steel. The distance it can be held away from the steel depends on the plate thickness. Note that hangers rubbing against the steel can cause squeaks. (See "Top Flange Hanger Spacing" illustration.)

11. **Face-mount hangers** can be used. Make sure that the hangers are tall enough to support the top flanges of the joists. Otherwise use web stiffeners. (See "Face-Mount Hangers" illustration.) Be sure to use the correct length and diameter of nail.

12. **The bottom flange cannot be cut or notched** except for a bird's mouth. At a bird's mouth, the flange cut should not overhang the edge of the top plate. (See "Bottom Flange I-Joist" illustration.)

13. Leave a ¹⁄₁₆" gap between I-joists and the supporting member when I-joists are placed in hangers. (See "Gap Between I-Joist and Support" illustration.)

14. The **top flange** can be notched or cut only over the top of the bearing and should not extend beyond the width of the bearing. (See "Top Flange I-Joists" illustration.)

15. **The web can have round or square holes.** Check the information provided with the I-joist package. Typically the center of the span requires the least strength and can have the biggest holes. The closer to the bearing point, the smaller the hole should be.

16. **When I-joists are used on sloped roofs,** they must be supported at the peak by a beam. This is different from dimensional lumber, where rafters may not require such a beam.

In working with residential I-joists, you should be aware that the APA has developed a standard for residential I-joists called Performance Rated I-joists (PRI). This standard shows the span and spacing for various uses for marked I-joists. (See "APA Performance Rated I-Joists" illustration.)

Solid Blocking and I-Joist Minimum Bearing

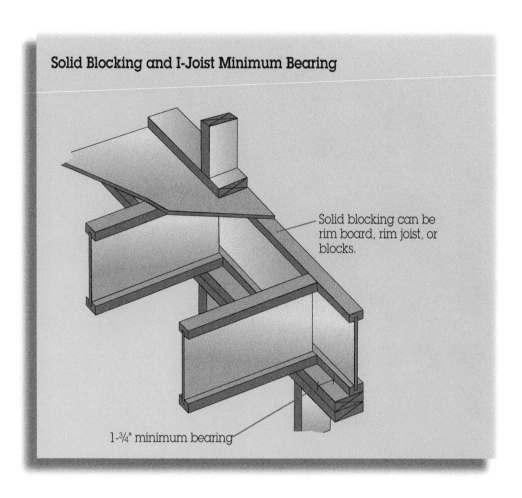

Solid blocking can be rim board, rim joist, or blocks.

1-¾" minimum bearing

Interior Bearing Wall Blocking Panel

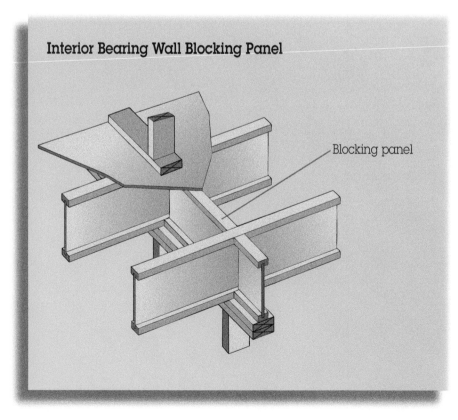

Blocking panel

Squash Blocks

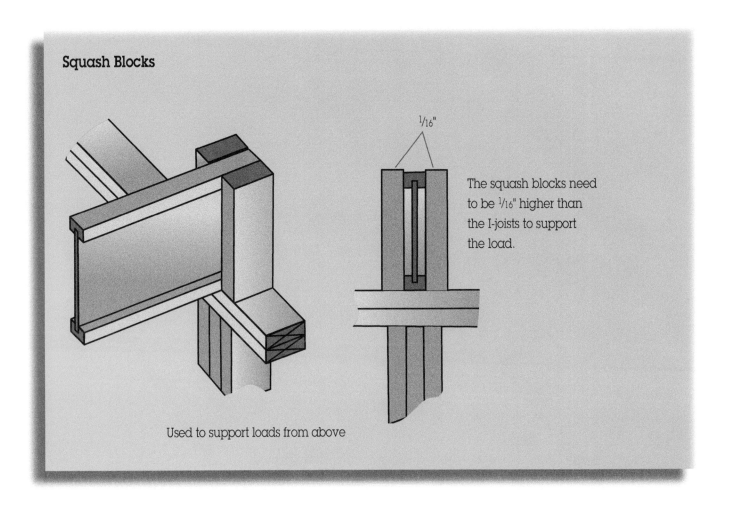

The squash blocks need to be $1/16$" higher than the I-joists to support the load.

Used to support loads from above

Web Stiffener

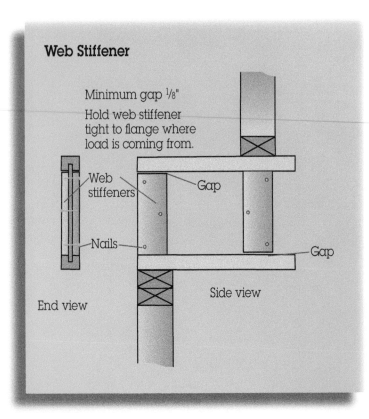

Minimum gap $1/8$"

Hold web stiffener tight to flange where load is coming from.

Web stiffeners

Gap

Nails

Gap

End view

Side view

Filler Blocking and Backer Blocking

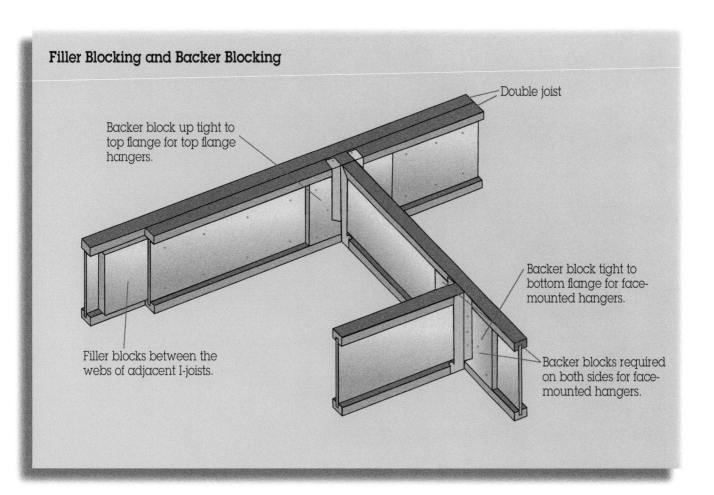

Double joist

Backer block up tight to top flange for top flange hangers.

Backer block tight to bottom flange for face-mounted hangers.

Filler blocks between the webs of adjacent I-joists.

Backer blocks required on both sides for face-mounted hangers.

Top Flange Hanger Tight

Hanger needs to be tight against header or backer block.

Use Right Size Nail

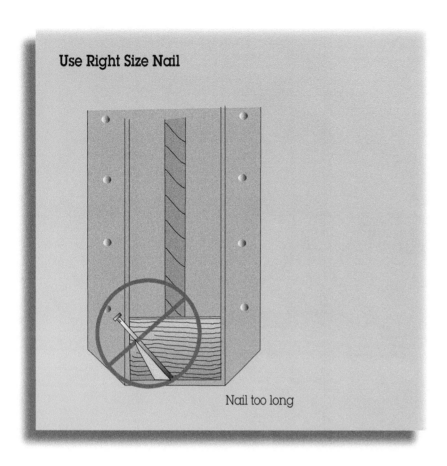

Nail too long

Top Flange Hanger Spacing

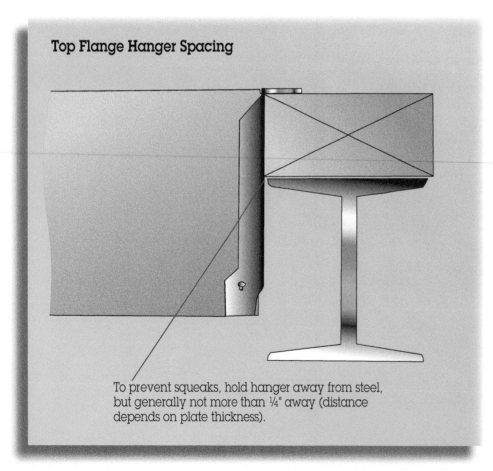

To prevent squeaks, hold hanger away from steel, but generally not more than ¼" away (distance depends on plate thickness).

Face-Mount Hangers

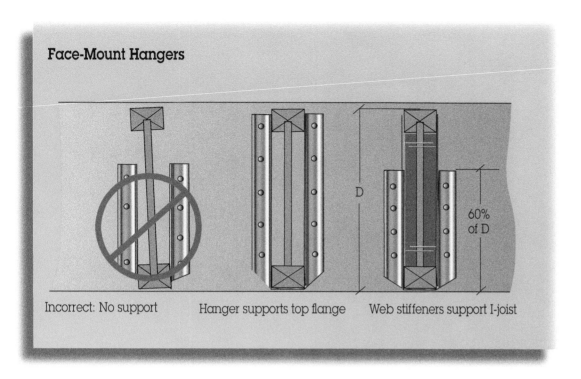

D

60% of D

Incorrect: No support

Hanger supports top flange

Web stiffeners support I-joist

Bottom Flange I-Joist

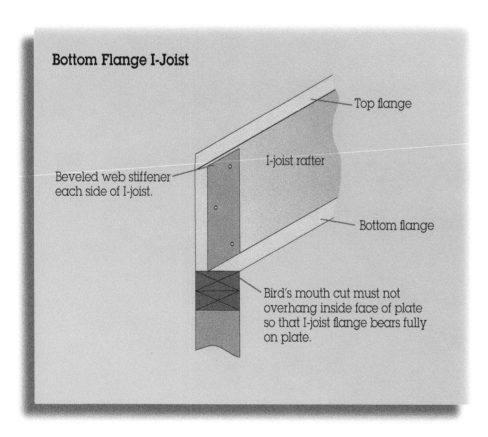

Top flange

I-joist rafter

Beveled web stiffener each side of I-joist.

Bottom flange

Bird's mouth cut must not overhang inside face of plate so that I-joist flange bears fully on plate.

Gap Between I-Joist and Support

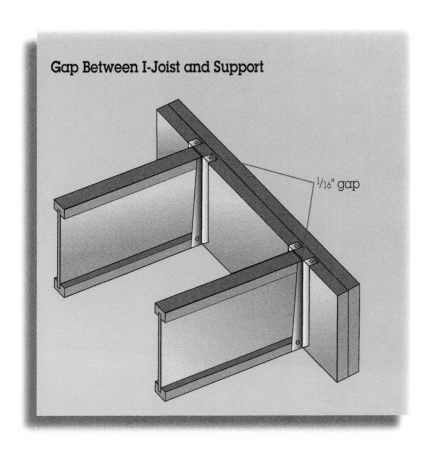

¹⁄₁₆" gap

Top Flange I-Joists

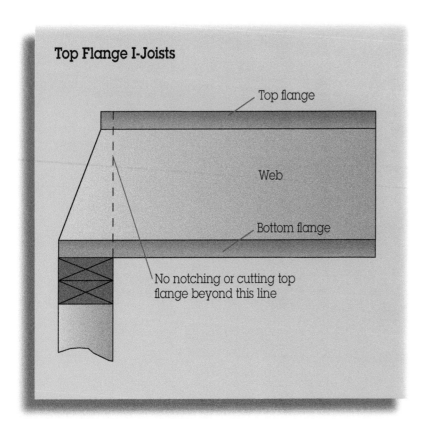

Top flange

Web

Bottom flange

No notching or cutting top
flange beyond this line

I-Joist Sizes

I-Joists	D (depth)	FW (flange width)
TJI 110	9-½", 11-⅞", 14", 16"	1-¾"
TJI 210	9-½", 11-⅞", 14", 16"	2-1/16"
TJI 230	9-½", 11-⅞", 14", 16"	2-5/16"
TJI 360	11-⅞", 14", 16", 18", 20"	2-5/16"
TJI 560	11-⅞", 14", 16", 18", 20"	3-½"
TJI 560D	18", 20", 22", 24"	3-½"
LPI 36	9-½", 11-⅞", 14"	2-¼"
LPI 18	9-½", 11-⅞", 14"	2-½"
LPI 20 Plus	9-½", 11-⅞", 14"	2-½"
LPI 32 Plus	9-½", 11-⅞", 14", 16"	2-½"
LPI 42 Plus	9-½", 11-⅞", 14", 16", 18", 20", 22", 24"	3-½"
LPI 52 Plus	11-⅞", 14", 16", 18", 20", 22", 24"	3-½"
LPI 56	9-½", 11-⅞", 14", 16", 18", 20", 22", 24"	3-½"
GPI 20	9-½", 11-⅞"	1-¾"
GPI 40	9-½", 11-⅞", 14"	2-5/16"
GPI 65	11-⅞", 14", 16"	2-7/16"
GPI 90	11-⅞", 14", 16"	3-½"
WI 40	9-¼", 16"	2-½"
WI 60	9-¼", 16"	2-½"
WI 80	9-¼", 16"	3-½"
RFPI 20	9-½", 11-⅞", 14"	1-¾"
RFPI 400	9-½", 11-⅞", 14", 16"	2-1/16"
RFPI 40	9-½", 11-⅞", 14", 16"	2-5/16"
RFPI 70	9-½", 11-⅞", 14", 16"	2-5/16"
RFPI 40S	9-½", 11-⅞", 14", 16"	2-½"
RFPI 60S	9-½" 11-⅞", 14", 16"	3-½"
RFPI 90	11-⅞", 14", 16"	3-½"
RFPI 80S	11-⅞", 14", 16"	3-½"
BCI 4500	9-½", 11-⅞", 14", 16"	1-¾"
BCI 5000	9-½", 11-⅞", 14", 16"	2"
BCI 6000	9-½", 11-⅞", 14", 16"	2-5/16"
BCI 60 2.0	11-⅞", 14", 16"	2-5/16"
BCI 6500 1.8	9-½", 11-⅞", 14", 16"	2-9/16"
BCI 90 2.0	11-⅞", 14", 16", 18", 20"	3-½"
AJS 5 to 200	9-¼" to 20"	2-½"
AJS 24, 25, 30	9-¼" to 24"	3-½"
Red-145	9-½", 11-⅞", 14"-16"	1-¾"
Red-165	11-⅞", 14"-30"	2-½"
Red-190	11-⅞", 14"-30"	3-½"
Red-190H	11-⅞", 14"-30"	3-½"
Red-190HS	11-⅞", 14"-32"	3-½"

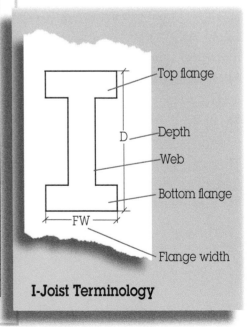

I-Joist Terminology

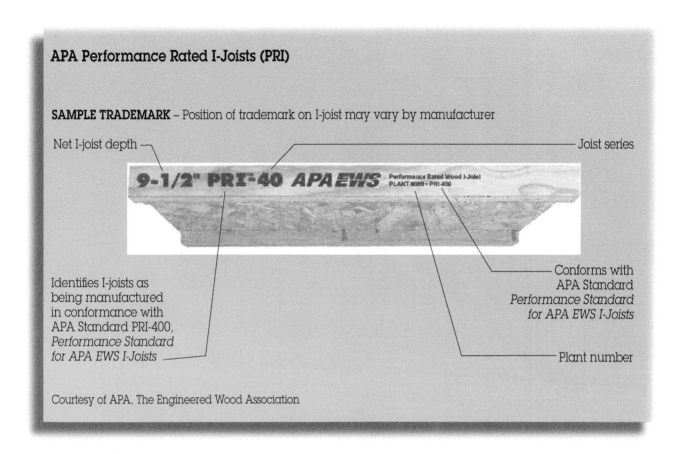

APA Performance Rated I-Joists (PRI)

SAMPLE TRADEMARK – Position of trademark on I-joist may vary by manufacturer

Net I-joist depth

Joist series

9-1/2" PRI-40 APA EWS

Performance Rated Wood I-Joist
PLANT #080 · PRI-400

Identifies I-joists as
being manufactured
in conformance with
APA Standard PRI-400,
*Performance Standard
for APA EWS I-Joists*

Conforms with
APA Standard
*Performance Standard
for APA EWS I-Joists*

Plant number

Courtesy of APA, The Engineered Wood Association

Glu-Lam Beams

Glu-lam beams are used when extra strength and greater spans are needed. They are usually big, heavy, and expensive, and require hoisting equipment to set them in place. Most often glu-lam beams are engineered for particular jobs. Glu-lam beams are produced by gluing certain grades of dimensional lumber together in a specific order. Many times the pieces are glued together to create a specific shape or camber. If a camber is created, the top of the beam will be marked. Make sure your crew installs it right-side-up.

The expense of glu-lam beams and the time required for replacing one makes it very important that they are cut correctly.

Notching and Drilling

The general rule for glu-lam beams is no notching or drilling without an engineer's direction. The engineer who determined the strength needed for the glu-lams is the person who will know how a notch or hole will affect the integrity of the glu-lam beam.

The way glu-lam beam connections are made will affect the strength and integrity of the beams. Following the illustrations are examples of correct and incorrect ways to connect glu-lam beams, and some tips for easy installation.

Cut Edge Full Bearing

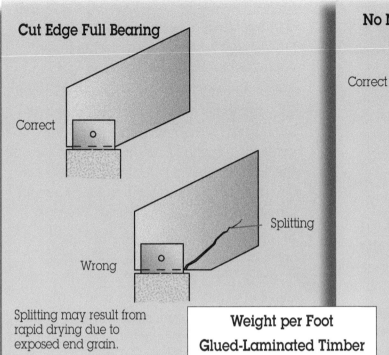

Correct

Wrong

Splitting

Splitting may result from rapid drying due to exposed end grain.

No Notching End of Glu-Lam Beam

Correct

Provide ½" minimum air space between wood and masonry surface.

Wrong

Splitting

Notching at ends of beams can cause splitting at inside corner. A notch at the end of a glu-lam beam should be checked by the engineer.

Weight per Foot
Glued-Laminated Timber Weights for DF-L

Width / Depth	3-⅛"	5-⅛"	6-¾"	8-¾"	10-¾"	12-¼"	14-¼"
6"	4.6	7.5	9.8	12.8	15.7	17.9	20.8
9"	6.8	1.2	14.8	19.1	23.5	26.8	31.2
12"	9.1	14.9	19.7	25.5	31.4	35.7	41.6
15"	11.4	18.7	24.6	31.9	39.2	44.7	52
18"	13.7	22.4	29.5	38.3	47	53.6	62.3
21"	16	26.2	34.5	44.7	54.9	62.5	72.7
24"	18.2	29.9	39.4	51	62.7	71.5	83.1
27"	20.5	33.6	44.3	57.4	70.6	80.4	93.5
30"	22.8	37.4	49.2	63.8	78.4	89.3	103.9
33"		41.1	54.2	70.2	86.2	98.3	114.3
36"		44.8	59.1	76.6	94.1	107.2	124.7
39"		48.6	64	83	101.9	116.1	135.1
42"		52.3	68.9	89.3	109.7	125.1	145.5
45"		56	73.8	95.7	117.6	134	155.9
48"		59.8	78.8	102.1	125.4	142.9	166.3
51"			83.7	108.5	133.3	151.9	176.7
54"			88.6	114.8	141.1	160.8	187
57"			93.5	121.2	148.9	169.7	197.4
60"			98.4	127.6	156.8	178.6	207.8
63"				134	164.6	187.6	218.2
66"				140.4	172.4	196.5	228.6
69"				146.8	180.3	205.5	239
72"				153.1	188.1	214.4	249.4
75"				159.5	196	223.3	259.8
78"					203.8	232.2	270.3
81"					211.7	241.2	280.5
84"					219.5	250.1	290.9
87"					227.3	259.1	301.4
90"					235.2	268.1	311.8
93"		For weights of HF multiply by .77				276.8	322
96"						285.8	332.5
99"		For weights of SP multiply by 1.03				294.8	343
102"						303.8	353.4
105"						312.6	363.6
108"		This weight chart is a good resource				321.6	374.1
111"		when you need to determine what size				330.6	384.5
114"		boom truck or crane you will need to					395
117"		set your glued-laminated beams					405.2
120"							415.6
123"							426.1
126"							436.5
129"							446.7

Tips for Installing Glu-Lam Beams

- For glu-lam beams that are installed at a pitch and need to have the bottom cut to be level, make sure that the end of the bottom cut closest to the bearing edge receives full bearing. (See "Cut Edge Full Bearing" illustration.)

- Ends of beams should not be notched unless approved by the engineer. (See "No Notching End of Glu-Lam Beam" illustration.)

- Glu-lam beams will shrink as they dry out. If the top of the beam is connected in a way that doesn't allow for shrinkage, the glu-lam beam will split. (See "Glu-lam Beam Shrinkage" illustration.)

- When a lateral support plate is used to connect two glu-lam beams, the holes should be slotted horizontally to prevent splitting. (See "Lateral Support Plate" illustration).

- Glu-lams are also used for posts. It is important to keep them away from concrete, which contributes to their decay. Placing a steel shim under the beam will keep it from touching the concrete. (See "Decay Prevention Next to Concrete" illustration.)

- Hinge connectors should be installed so that they don't cause splitting of the glu-lam beams. This can be done by using a strap that is independent of the hinge connector, or by vertical slotting the holes in a strap that is connected to the hinge connector. (See "Hinge Connector Slotted Holes" illustration.)

- Glu-lam beams rest on metal post caps that often have a weld or radius in the bottom corner. If you don't ease the bottom corners of the beam, the beam will sit up in the pocket. Often, the glu-lam beam's bottom corners are already rounded and won't need attention.

- In some cases, the sides of the metal post caps are bent in so that the beams will not slide in properly. Check all the sides of the metal post caps before they are installed, so you won't have a forklift or boom truck and crew standing around waiting while someone labors on top of a ladder to widen the sides of the post cap. (See "Forklift setting glu-lam beams" photograph later in this chapter.)

- Glu-lam beams are often attached to metal caps with bolts. The holes can be drilled either before or after setting the glu-lam beams. If the holes are drilled after the beams are set, use a drill with a clutch. It's easy to break a wrist or get thrown from a ladder when a ½" drill motor without a clutch gets caught on the metal.

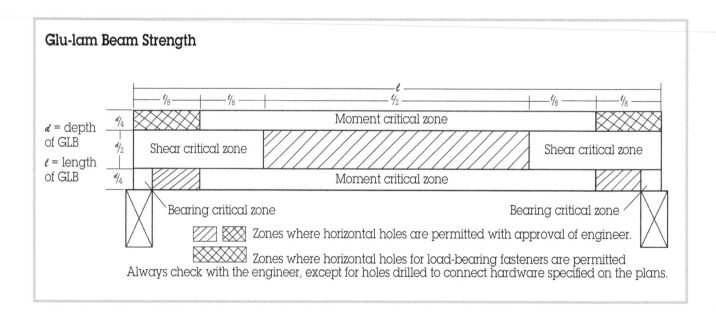

Glu-lam Beam Strength

d = depth of GLB

ℓ = length of GLB

Moment critical zone · Shear critical zone · Bearing critical zone

Zones where horizontal holes are permitted with approval of engineer.

Zones where horizontal holes for load-bearing fasteners are permitted

Always check with the engineer, except for holes drilled to connect hardware specified on the plans.

Glu-lam Beam Shrinkage

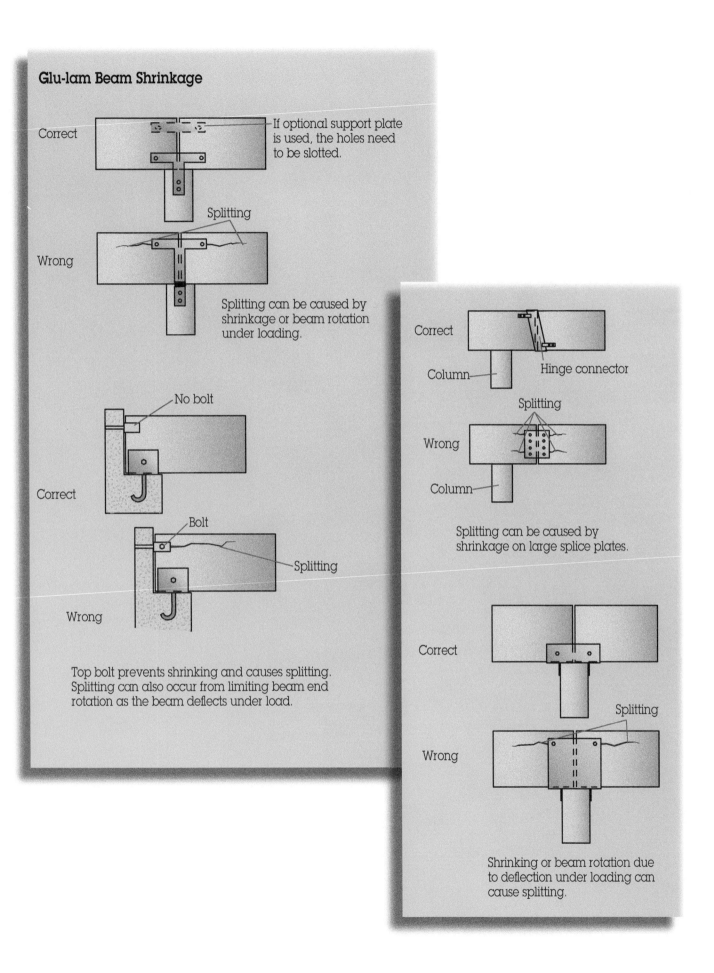

Correct

If optional support plate is used, the holes need to be slotted.

Splitting

Wrong

Splitting can be caused by shrinkage or beam rotation under loading.

No bolt

Correct

Bolt

Splitting

Wrong

Top bolt prevents shrinking and causes splitting. Splitting can also occur from limiting beam end rotation as the beam deflects under load.

Correct

Column

Hinge connector

Splitting

Wrong

Column

Splitting can be caused by shrinkage on large splice plates.

Correct

Splitting

Wrong

Shrinking or beam rotation due to deflection under loading can cause splitting.

177

Lateral Support Plate

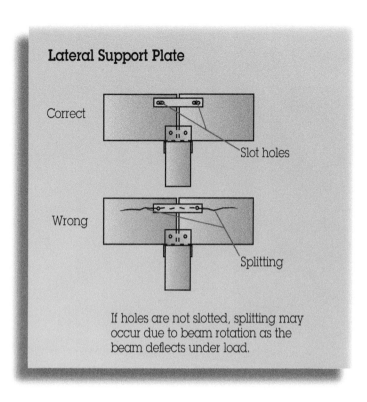

Correct

Slot holes

Wrong

Splitting

If holes are not slotted, splitting may occur due to beam rotation as the beam deflects under load.

Decay Prevention Next to Concrete

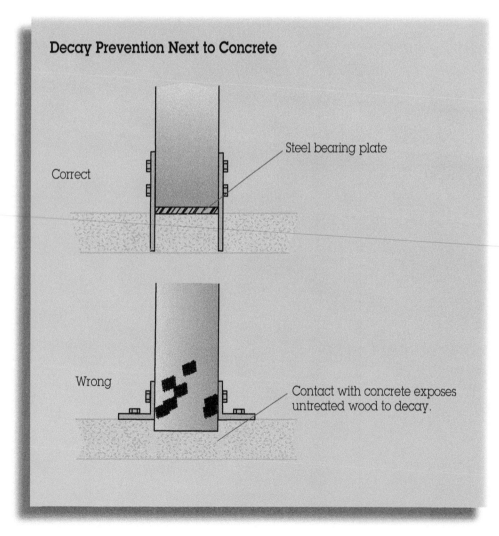

Steel bearing plate

Correct

Wrong

Contact with concrete exposes untreated wood to decay.

Hinge Connector Slotted Holes

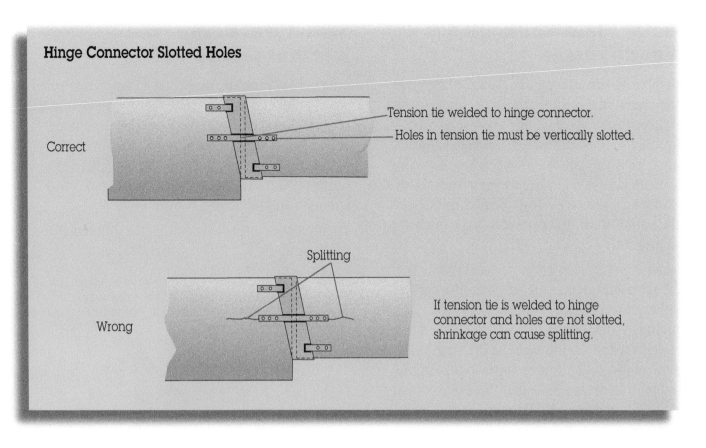

Correct

Tension tie welded to hinge connector.

Holes in tension tie must be vertically slotted.

Splitting

Wrong

If tension tie is welded to hinge connector and holes are not slotted, shrinkage can cause splitting.

Hinge Connectors

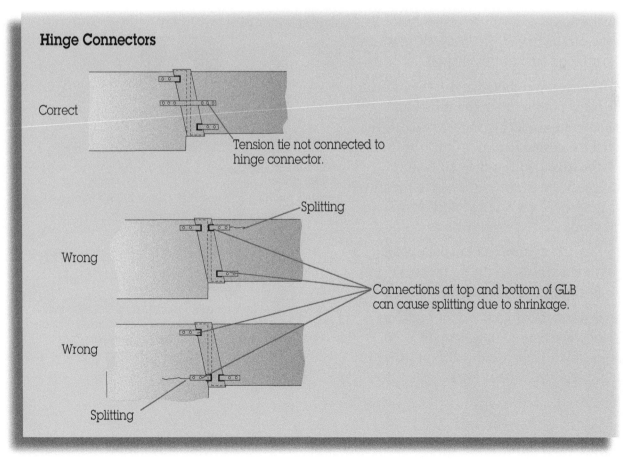

Correct

Tension tie not connected to hinge connector.

Splitting

Wrong

Connections at top and bottom of GLB can cause splitting due to shrinkage.

Wrong

Splitting

Metal Plate-Connected Wood Trusses

Metal plate-connected (MPC) wood trusses were first used in the early 1950s. Today they are used in more than 75% of all new residential roofs. Basically they are dimension lumber engineered and connected with metal plates. Less expensive than alternative roof systems, these trusses can also span longer distances. The "Pitched Truss Parts" illustration shows the parts of a single pitched truss on page 182.

Because MPC trusses are engineered products, they should never be cut, notched, spliced, or drilled without first checking with the designing engineer. Building codes require that a truss design drawing be delivered to the job site. The drawings must show, among other things, the layout locations and bracing details. Note that these drawings are typically not made with framers in mind, so it might take some study time to figure out where the engineer wants the braces. The bracing details often show the braces as small rectangles running laterally between the trusses. See "Lateral Truss Bracing" illustration later in this chapter.

When flying trusses, you should attach the cables around the panel points. When the trusses are greater than 30', a spreader bar should be used. The cables should toe inward to prevent the truss from buckling. If the truss is longer than 60', you will need a strongback temporarily attached to the truss to stabilize it. (See "Flying Trusses" illustration, later in this chapter.)

If you have multiple trusses, you can build a sub-assembly of several trusses on the ground with cross braces and sheathing, then erect them together.

When trusses sit on the ground, on the building, or in place for any length of time, keep them as straight as possible. They are more difficult to set in place and to straighten if they have not been stored properly on site.

Structural Composite Lumber (SCL)

Structural composite lumber (SCL) is an engineered wood product that combines veneer sheets, strands, or small wood elements with exterior structural adhesives. The most common of these products are laminated veneer lumber (LVL), parallel strand lumber (PSL), laminated strand lumber (LSL) and Oriented Strand Lumber (OSL).

Their names pretty well describe the differences between them.

Laminated Veneer Lumber (LVL) is made by bonding thin wood veneers together in a large billit so the grain of the veneers is parallel in the long direction. The LVL billet is sawn to desired length and width.

Parallel Strand Lumber (PSL) is made from veneers chipped into long strands and laid in parallel form and bonded together with an adhesive.

Laminated Strand Lumber (LSL) is made from flaked wood strands. The strands are combined with an adhesive and oriented and formed into a large mat or billet and pressed.

Oriented Strand Lumber (OSL) is made from flaked wood strands. The strands are combined with an adhesive, oriented and formed into a large mat or billet and pressed.

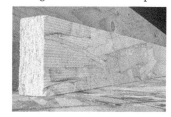

Cross-Laminated Timber (CLT) is prefabricated, solid engineered wood panels. It is made with kiln-dried lumber boards or structural laminated lumber. It is laminated in alternating directions and bonded with structural adhesives that form a solid, straight, regular panel. The panels are made with odd number layers, usually three to seven. The typical width is 2 feet, 4 feet, 8 feet, and 10 feet and a thickness of 20 inches or less with a length of up to 60 feet.

Like other engineered products, structural composite lumber requires that you follow the engineered specifications that will appear on the plans.

Sometimes the specifications simply indicate the use of a particular piece of SCL in a particular location. For larger jobs, you will find the SCL requirements called out in the shop drawings or the structural plans.

Because these are engineered products, you must consult the design engineer before you can drill or notch. Some manufacturers provide guidelines for drilling and notching, but this is not typical.

SCL has the advantages of dimensional consistency, stability, and availability of various sizes. It is important to note, however, that where dimensional lumber 4 × 10s, 4 × 12s, etc. can shrink significantly, SCLs have minimal shrinkage. The engineer should allow for this in the design so that you will not have to consider this factor when using SCLs as the plans specify.

Note that SCL studs are becoming common in building tall walls. They provide a degree of straightness that dimensional lumber does not. Although they are heavy and, as a result, not so easy to work with, they make nice, straight walls.

Conclusion

Engineered wood products come in a variety of forms. Becoming familiar with these products is important if you plan to work with them. Always be sure to follow manufacturers' directions, and always consult an engineer if you plan to cut, notch, or drill engineered wood product components.

Forklift setting glu-lam beams

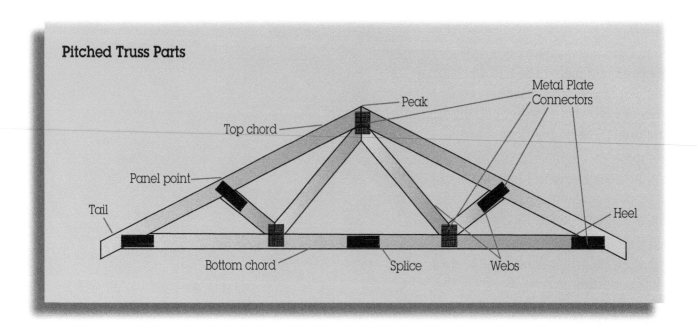

Pitched Truss Parts

Peak

Metal Plate
Connectors

Top chord

Panel point

Tail

Heel

Bottom chord

Splice

Webs

Lateral Truss Bracing

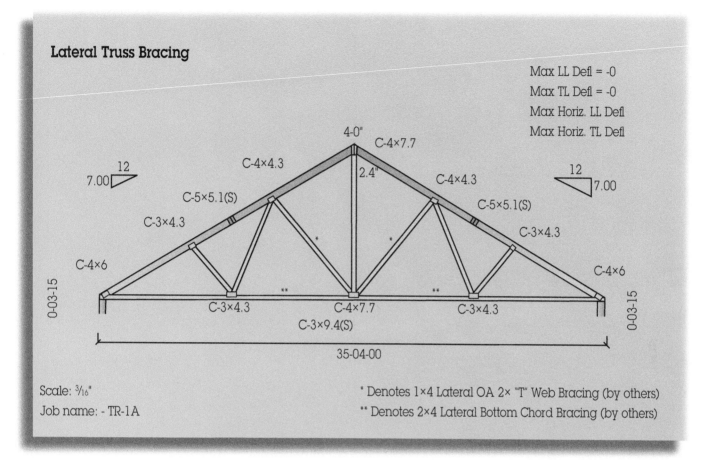

Max LL Defl = -0
Max TL Defl = -0
Max Horiz. LL Defl
Max Horiz. TL Defl

4-0"
C-4×7.7
C-4×4.3
2.4"
C-4×4.3
12
7.00
C-5×5.1(S)
C-5×5.1(S)
12
7.00
C-3×4.3
C-3×4.3
C-4×6
C-4×6
0-03-15
0-03-15
C-3×4.3
C-4×7.7
C-3×4.3
C-3×9.4(S)
35-04-00

Scale: ³/₁₆"
Job name: - TR-1A

* Denotes 1×4 Lateral OA 2× "T" Web Bracing (by others)
** Denotes 2×4 Lateral Bottom Chord Bracing (by others)

Flying Trusses

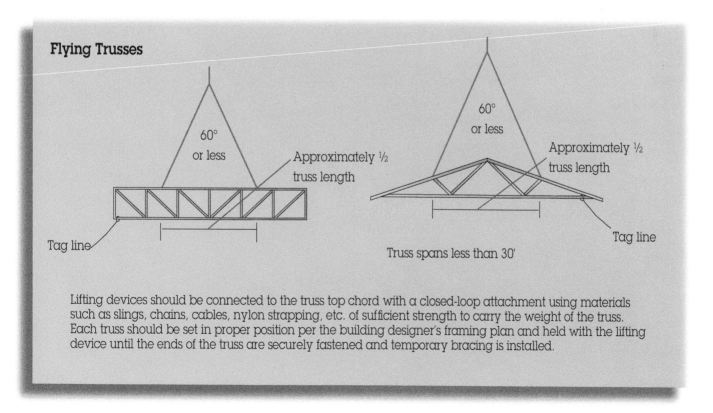

60°
or less

Approximately ½
truss length

60°
or less

Approximately ½
truss length

Tag line

Tag line

Truss spans less than 30'

Lifting devices should be connected to the truss top chord with a closed-loop attachment using materials such as slings, chains, cables, nylon strapping, etc. of sufficient strength to carry the weight of the truss. Each truss should be set in proper position per the building designer's framing plan and held with the lifting device until the ends of the truss are securely fastened and temporary bracing is installed.

Flying Trusses (continued)

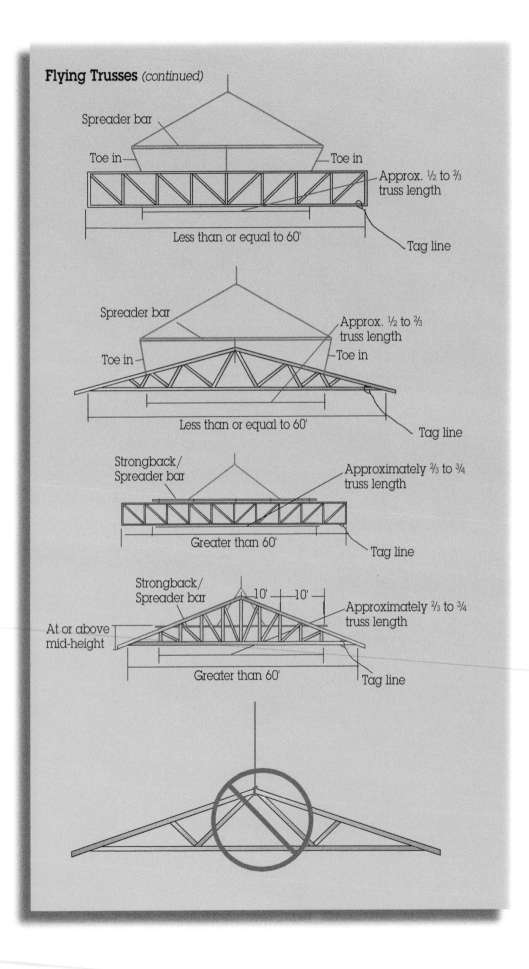

Spreader bar
Toe in
Toe in
Approx. ½ to ⅔ truss length
Tag line
Less than or equal to 60'

Spreader bar
Approx. ½ to ⅔ truss length
Toe in
Toe in
Less than or equal to 60'
Tag line

Strongback/ Spreader bar
Approximately ⅔ to ¾ truss length
Greater than 60'
Tag line

Strongback/ Spreader bar
10'
10'
Approximately ⅔ to ¾ truss length
At or above mid-height
Greater than 60'
Tag line

184

Chapter Nine
WIND AND EARTHQUAKE FRAMING

Contents

Chapter Nine

WIND AND EARTHQUAKE FRAMING

Buildings are naturally affected by the forces of nature, and also by artificial forces. Elements such as gravity, wind, snow, earthquakes, retained soil, water, impact by an object, and mudslides can all have negative effects on a building.

This chapter will give you a basic understanding of the forces that affect buildings, and some helpful information on the framing methods used to resist those forces.

Although you may not be responsible for designing structural requirements for buildings, it is important to have some understanding of a building's structural loads. When you are aware of the reasons behind the decisions engineers and architects make, it is easier to interpret the plans, and to make sure that the structure is built accordingly.

The Strength of Good Framing

The forces of nature can have devastating effects on buildings. The following photo shows an example of how destructive the elements can be. This photo is quite dramatic; you can see that the ground literally fell out from under the house. But the photo also shows the strength of good framing—the house stayed together even though the ground collapsed under it.

Source: APA, The Engineered Wood Association

The house stayed together even as the ground fell from under it.

Understanding Structural Loads

As the forces of nature contact a building, they travel throughout seeking a weak link. Ultimately, if a weak link is not found, the force or energy will be transferred to the ground, which will absorb the force. Each component of the building must be strong enough to transfer the load in a path to the ground. The components are:

- Foundations
- Walls
- Floors
- Roofs
- Connections

To achieve the strength needed, a building's walls, floors, and roof must work together as a unit. The vertical elements that are used to resist forces are commonly called **shear walls,** and the horizontal elements (like floors and roofs) are called **diaphragms.** The path of energy to the ground is called the **load path.** The diagram on the next page shows the load path for transferring the forces to the ground.

Building Code Load Requirements

Conventional and *nonconventional* codes regulate the strength needed in the walls, floors, roofs, and connections to resist the forces on buildings. The conventional code describes a *prescriptive* standard to resist the forces. The standard applies to simple buildings using common construction methods. The nonconventional code is a performance-rated system and provides *non-prescriptive* engineering guidelines that can be applied to more unusual or more difficult buildings.

Prescriptive Format

The prescriptive format has specific requirements, such as the size of studs needed or the type of wall bracing. If you build the structure following these requirements, then the building meets the minimum code standards for a safe building. The prescriptive codes are covered in more detail in Chapter 10.

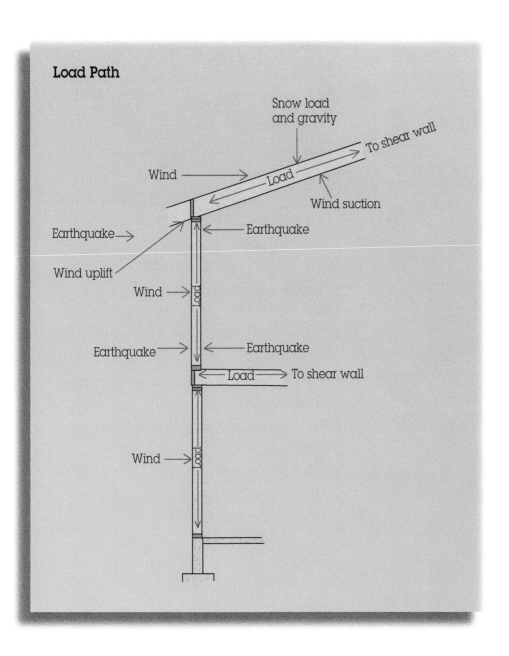

Load Path

189

Framers meet prescriptive code requirements on a regular basis, sometimes without even knowing it. As they brace their walls, block and nail their floor system, nail their walls to the floors, and bolt the building to the foundation, they are creating a load path that transfers the forces of nature to the ground—in ways that are prescribed by the code.

Non-Prescriptive Code

The performance, or non-prescriptive, code provides for free design, as long as it stays within certain code standards. Performance designing is different for each building, and the engineer or architect must specify and detail all aspects of the design.

A special design might be needed because a building is in a high-earthquake or a high-wind zone, because it requires large open spaces or window walls, or to resist other forces. The most common forces affecting buildings are shown in the illustration "Forces on Buildings."

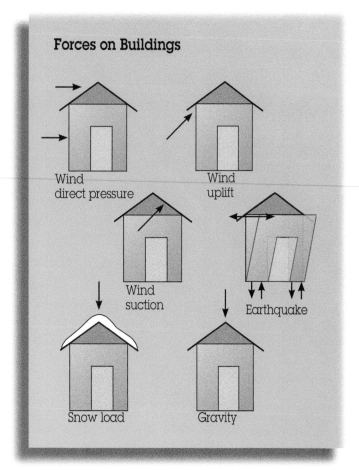

Forces on Buildings

Wind direct pressure

Wind uplift

Wind suction

Earthquake

Snow load

Gravity

Regional Considerations

Different forces affect buildings in the various parts of the country. Builders have to worry about earthquakes in California, high winds in Florida, and snow loads in Colorado. It's easier to understand the architect's or engineer's plans if you are aware of these factors. The following maps give you an idea of some of the areas of the country that suffer most from the effects of earthquakes, winds, and snow loads.

Framing Details

The most common framing details can be broken down into three categories.

- Shear wall construction
- Diaphragm construction
- Connections

Each of these categories is covered in this section, including important points for framing.

Shear Wall Construction

The factors that affect the strength of any shear wall are:

- The size and type of material used for the plates and studs.
- The size and type of material used for the sheathing.
- Whether one side or both sides have sheathing.
- The nail sizes and patterns.
- Whether or not there is blocking for all the edges of the sheathing.

Engineers and architects are free to use any system they prefer, as long as they can prove that it meets the minimum strength requirements.

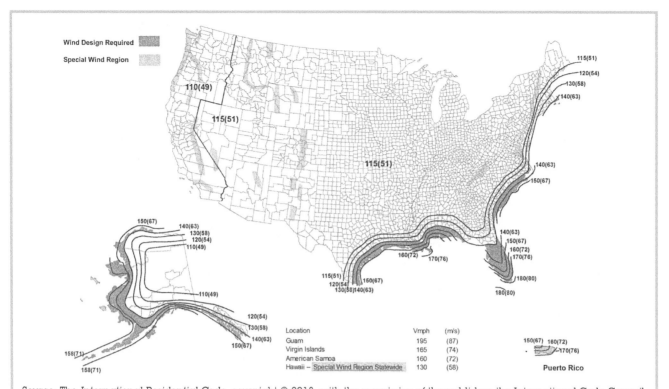

Source: The *International Residential Code*, copyright © 2018, with the permission of the publisher, the International Code Council. The 2018 *International Residential Code* is a copyrighted work of the International Code Council.

Wind speeds correspond to approximately a 7% probability of exceedance in 50 years

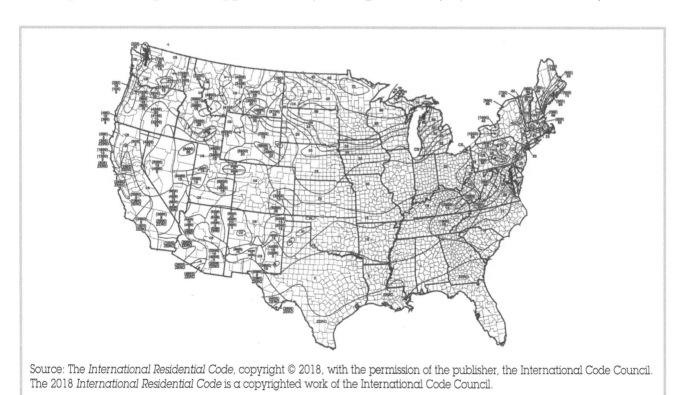

Source: The *International Residential Code*, copyright © 2018, with the permission of the publisher, the International Code Council. The 2018 *International Residential Code* is a copyrighted work of the International Code Council.

Ground snow loads, P_g for the United Stations (lb/ft²)

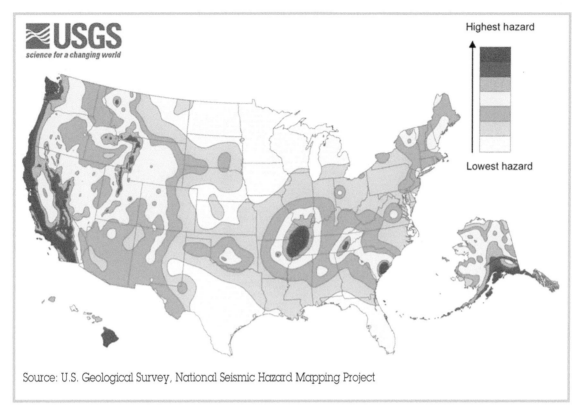

Source: U.S. Geological Survey, National Seismic Hazard Mapping Project

Seismic Map of Continental U.S.

If there are many shear walls in a building, the engineer usually creates a schedule from a code table to show the wall requirements. Unfortunately, there is no standard for labeling shear walls, so the schedules made by the engineers may all be different. They do, however, usually have common components. You will need to study the shear wall schedule on the plans to understand all the components that apply to framing.

Refer to the Shear Wall Schedule table later in this chapter for an example. It is an easy one to use because the labels also identify the nailing pattern and the type of sheathing. It was developed by the framing council in the state of Washington.

Important Points for Shear Wall Framing

1. **Stud sizes**—Specified nailing patterns may require changes in the stud sizes. There are three conditions where 3× studs are required for nailing adjoining sheathing edges. A fourth condition is required in seismic design category D, E, or F.

   - If the edge nailing is 2-½" O.C. or less.

   - If there is sheathing on both sides of the wall, the adjoining sheathing edges fall on the same stud on both sides of the wall, and the nailing pattern is less than 6" O.C.

   - If 10d (3" × 0.148") nails are used with more than 1-½" penetration, and they are spaced 3" or less O.C.

   - (For seismic design categories D, E, or F) where shear design values exceed 350 pounds per linear foot.

2. **Penetration**—It is very important that the nail does not penetrate the outside veneer of the sheathing (see "Nail Penetration" illustration.) A pressure regulator or nail-depth gage can be used to make sure this doesn't happen. (See "Nail regulator and flush nailer" illustration.) The top of the nail should be flush with the surface of the sheathing.

3. **Nail size**—The nail size may change from wall to wall. Check the specified thickness and length of the nails.

4. **Nail spacing**—The pattern for nailing the sheathing to the intermediate framing members is usually the standard 12" on center. It is the edge nailing that changes to increase the strength. If 3× studs are required, then the pattern must be staggered. Make sure that the nails are at least ⅜" away from the edge of the sheathing.

5. **Blocking**—The details or shear wall schedule should specify whether blocking is required for panel edges. If the wall is 8' or less in height, you can usually satisfy this requirement by running the plywood vertically, so that all the panel edges have backing.

Diaphragm Construction

The strength of diaphragms is affected by these factors:

- The size and type of material used for the joists or rafters

- The size and type of material used for the sheathing

- The direction of the sheathing in relation to the members it is attached to

- The nail sizes and patterns

- Any blocks, bridging, or stiffeners

Framer-Friendly Tips

3× studs are used in shear walls at adjoining edges where the nail spacings are small. Check the shear wall schedule.

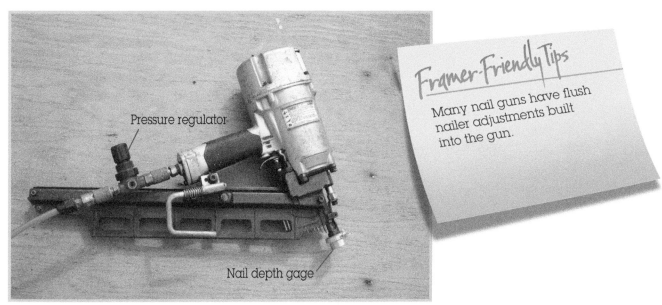

Framer-Friendly Tips

Many nail guns have flush nailer adjustments built into the gun.

Pressure regulator

Nail depth gage

Nail regulator and flush nailer shown affixed to a pneumatic nailer

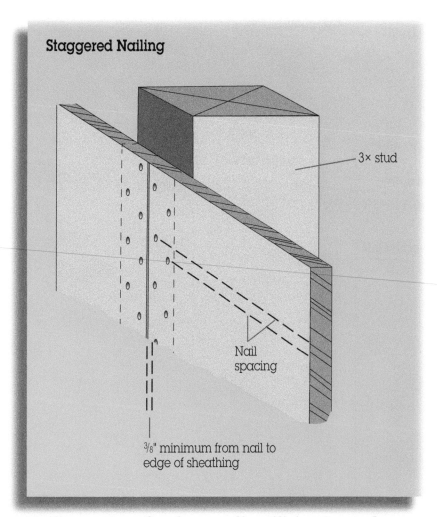

Staggered Nailing

3× stud

Nail spacing

³⁄₈" minimum from nail to edge of sheathing

Nailing pattern for shear walls utilizing 3× studs

Building codes provide tables for diaphragms similar to those for shear walls. To summarize, the variables used to increase the strength of the diaphragm are: the thickness of the sheathing, the size of the nails, the width of the framing member, the nail spacing, and whether or not the diaphragm is blocked.

Diaphragm Framing Tasks of Particular Concern:

- **Nail spacing**—The nailing pattern for nailing the sheathing to the intermediate framing members is usually the standard 12" O.C. It is the edge nailing that changes to increase the strength.
- **Penetration**—The nail must not penetrate the sheathing's outside veneer.

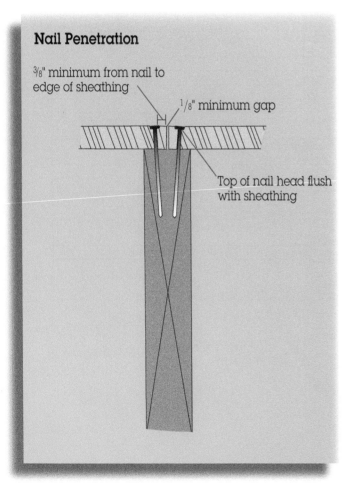

Nail Penetration

3/8" minimum from nail to edge of sheathing

1/8" minimum gap

Top of nail head flush with sheathing

- **Nail size**—The nail sizes will vary based on the engineer's design, or code requirements. Check the specified thickness and length.
- **Blocking**—It is common to have blocking in the joist space that runs parallel to the exterior walls. It will be detailed on the plans if it is required. Blocking can also be used on the edges of the sheathing.

Connections

"Connectors" can refer to beams or other construction elements, but in most cases, connectors are hardware specifically designed for common framing connections. As part of the load path, connections have to be strong enough to transfer the forces of nature.

In the prescriptive code, the connections are made with anchor bolts to the foundation, and with nails to connect floor joists to the plates below them, wall bottom plates to floors, and rafters or trusses to wall plates.

In non-prescriptive design, there are many ways to achieve the required force transfer between the shear walls, diaphragms, and foundation. The most common method involves metal connectors, which are produced by many companies. The Simpson Strong-Tie Company, because of its work in developing, testing, and cataloging connectors, is often referenced in building plans. Simpson Strong-Tie connector catalog numbers will be used in the balance of this book. Please note that substitutes with equivalent strength are available.

Shear Wall Schedule

Label	APA Rated Sheathing [1][2][4][12][13]	Nail Size & Spacing @ Edges [4][5]	Stud & Blocking Size @ Adjoining Edges [3][6][14]	Rim Joist or Block Connection to Top Plate [7][8]	2 × Bottom Plate Attachment — Nailing To Wood Below [9]	Sill Plate Attachment — Anchor Bolt to Concrete Below [10][15]	Sill Plate Size @ Foundation [11]	PLF Capacity
W6	$^{15}/_{32}$" one side	0.131×2-½ @ 6" O.C.	2×	Clip @ 24" O.C.	.148×3-¼" @ 6"O.C.	⅝"@ 48" O.C.	2×	
W4								
W3								
W2								
2W4 [2]								
2W3 [2]	$^{15}/_{32}$" two sides	0.131×2-½ @ 3"O.C.	3×	Clip @ 12" O.C. EACH SIDE	Clip @ 12" O.C. EACH SIDE [7],[8]	⅝" @ 16" O.C.	3×	
2W2 [2]								

Required Notes

[1] Install panels either horizontally or vertically

[2] Where sheathing is applied on both sides of wall, panel edge joints on 2× framing shall be staggered so that joints on the opposite sides are not located on the same studs.

[3] Blocking is required at all panel edges.

[4] Provide shear wall sheathing and nailing for the entire length of the walls indicated on the plans. Ends of full height walls are designated by exterior of the building, corridors, windows, or doorways, or as designated on plans. See plans for holdown requirements. (Alternate note: walls designated as perforated shear walls require sheathing above and below all openings.)

[5] Sheathing edge nailing is required at all holdown posts. Edge nailing may also be required to each stud used in built-up holdown posts. Refer to the holdown details for additional information.

[6] Intermediate framing to be with 2× minimum members. Field nailing 12" O.C.

[7] Based on 0.131 × 1-½" long nails used to attach framing clips directly to framing. Use 0.131 × 2-½" nails where installed over sheathing.

[8] Framing clips: A35 or LTP5 or approved equivalent

[9] Where plate attachment specifies (2) rows of nails, provide double joist, rim or equal. Attach per details.

[10] (In Seismic Design Categories D, E, & F) Anchor bolts shall be provided with steel plate washers $^{3}/_{16}$"×2"×2". Embed anchor bolts 7" minimum into the concrete.

[11] Pressure-treated material can cause excessive corrosion in the fasteners. Provide hot-dipped galvanized (electro-plating is not acceptable) nails and connector plates (framing angles, etc.) for all connectors in contact with pressure treated framing members.

Alternate Notes

[12] $^{7}/_{16}$" APA rated sheathing (OSB) may be used in place of $^{15}/_{32}$" sheathing provided that all studs are spaced at 16" O.C.

[13] Where wood sheathing (W) is applied over gypsum sheathing (G), contact the engineer of record for alternate nailing requirements.

[14] At adjoining panel edges, (2) 2× studs nailed together may be used in place of a single 3× stud. Double 2× studs may be connected together by nailing the studs together with 3" long nails of the same spacing and diameter as the plate nailing.

[15] Contact the engineer of record for adhesive or expansion bolt alternatives to cast-in-place anchor bolts. (Special inspection may be required.)

There are connectors made for just about every type of connection you can think of. As the framer in charge, however, it is not your job to decide on the type of connector, but rather to use correctly the connector that is specified. The best way to do this is to read the specifications in the connector catalog. Following is an illustration from a Simpson Strong-Tie Catalog, and a good example of instructions for installing hold-downs. You can reference the connectors at www.strongtie.com.

There are different connectors for the variety of different framing details, but only four common areas of connection:

- Foundation
- Wall-to-wall
- Roof-to-wall
- Foundation-to-top-of-the-top-wall

Important Points for Connection Framing

- Install all connectors per catalog instructions.
- Drill holes no more than $\frac{1}{16}$" bigger than bolts.
- Use washers next to wood.
- Fill all nail holes unless using catalog specifications.
- Know that the connection is only as strong as the weakest side. Make sure to space and nail each side the same. (See "Equal Nailing" illustration later in this chapter.)
- Be aware that some connectors have different-shaped nail holes. The different-shaped holes have different meaning, as illustrated in "Nail Hole Shapes" later in this chapter.

Hold-Downs

Hold-downs are connections commonly used for foundations, wall-to-wall connections, wall-to-concrete connections, and wall- or floor-to-drag strut. Hold-downs are also called *anchor downs* and *tie-downs*. They can be difficult to install, but if you plan ahead and install as you go, the job is more manageable. Hold-downs that attach walls to the concrete foundation are typically attached to bolts already in the concrete. These bolts are generally set in place by the foundation crew. Sometimes they won't be set in the right place.

You will want to locate the hold-down as close to the end of the shear wall as possible. If the bolt is already in the concrete, you will have to locate a hold-down on either side of the bolt. When considering the location, be aware of how it relates to what is on the floor above it; you don't want, for example, the hold-down coming up in a door or window. You should also allow enough space to install and tighten nuts and bolts.

When to Install Hold-Downs

Although it is common to wait until the building is framed to install the hold-downs, waiting can also present problems, such as studs that are already nailed in place where you want to install the hold-downs, sheathing that is hard to nail because it may be on the exterior of a second or higher floor, and possible pipes or wires running in the stud cavity.

It is helpful to install the hold-down studs as you build the walls. The layout framer should detail the hold-down studs while detailing the wall plates, and should also drill the plates for the anchor bolt or the threaded rods. If an upper floor is involved, the framer should also drill down through the subfloor sheathing and the top and double plate of the wall on the floor below. The wall builder should drill the

This is a sample page from a hardware catalog:

HDQ8/HHDQ

SIMPSON
Strong-Tie

Holdowns

The HHDQ series of holdowns combines low deflection and high loads with ease of installation. The unique seat design of the HDQ8 greatly minimizes deflection under load. Both styles of holdown employ the Simpson Strong-Tie® Strong-Drive® SDS Heavy-Duty Connector screws which install easily, reduce fastener slip and provide a greater net section when compared to bolts. They may be installed either flush or raised off the mudsill without a reduction in load value.

Special Features:

- Uses Strong-Drive SDS Heavy-Duty Connector screws which install easily, reduce fastener slip, and provide a greater net section area of the post compared to bolts
- Strong-Drive SDS Heavy-Duty Connector screws are supplied with the holdowns to ensure proper fasteners are used
- No stud bolts to countersink at openings

Material: HDQ8 — 7 gauge; HHDQ — Body: 7 gauge, washer: ½" plate

Finish: HDQ8 — Galvanized; HHDQ — Simpson Strong-Tie® gray paint; HHDQ11 — Available in stainless steel

Installation:

- See General Notes on pp. 75–76
- No additional washer is required
- Strong-Drive SDS Heavy-Duty Connector screws install best with a low-speed high-torque drill with a ⅜" hex-head driver

HDQ8:

- ⅝" of adjustability perpendicular to the wall

HHDQ11/14:

- No additional washer is required
- HHDQ14 requires a heavy-hex anchor nut (supplied with holdown)

Codes: See p. 14 for Code Reference Key Chart

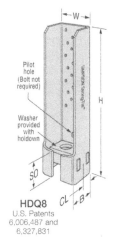

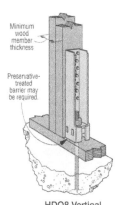

HDQ8
U.S. Patents
6,006,487 and
6,327,831

HDQ8 Vertical Installation

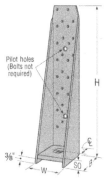

HHDQ11
(HHDQ14 similar)

Vertical HHDQ11 Installation
(HHDQ14 similar)

Not sure you have the right holdown?
Our Holdown Selector software is a great tool to help you select the best product for the job. Visit **strongtie.com/software**.

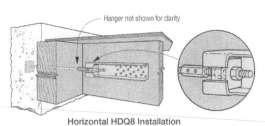

Horizontal HDQ8 Installation

Holdowns and Tension Ties

These products are available with additional corrosion protection. For more information, see p. 18.

Model No.	Ga.	Dimensions (in.)					Fasteners		Minimum Wood Member Thickness (in.)	Allowable Tension Loads (160)			Code Ref.
		W	H	B	CL	SO	Anchor Bolt Dia. (in.)	SDS Screws		DF/SP	SPF/HF	Deflection at Allowable Load (in.)	
HDQ8-SDS3	7	2⅞	14	2½	1¼	2⅜	⅞	(20) ¼" x 3"	3	5,715	4,115	0.064	I6, L8, FL
								(20) ¼" x 3"	3½	7,630	5,495	0.094	
								(20) ¼" x 3"	4½	9,230	6,645	0.095	
HHDQ11-SDS2.5	7	3	15⅛	3½	1½	⅞	1	(24) ¼" x 2½"	5½	11,810	8,505	0.131	
HHDQ14-SDS2.5	7	3	18¾	3½	1½	⅞	1	(30) ¼" x 2½"	7¼	13,015	9,370	0.107	
									5½²	13,710	10,745	0.107	

1. See pp. 75–76 for Holdown and Tension Tie General Notes.
2. Noted HHDQ14 allowable loads are based on a 5½"-wide post (6x6 min.). Other loads based on 3½"-wide post minimum.
3. HHDQ14 requires heavy-hex anchor nut (supplied with holdown).
4. HDQ and HHDQ installed horizontally achieve compression loads with the addition of a standard nut on the underside of the load transfer plate. Refer to ICC-ES ESR 2320 for design values. HDQ8 requires a standard nut and BP⅞-2 (sold separately) load washer on the underside of the holdown for compression load. Design of anchorage rods for compression force shall be per the Designer.

C-C-2017 ©2017 SIMPSON STRONG-TIE COMPANY INC.

Equal Nailing

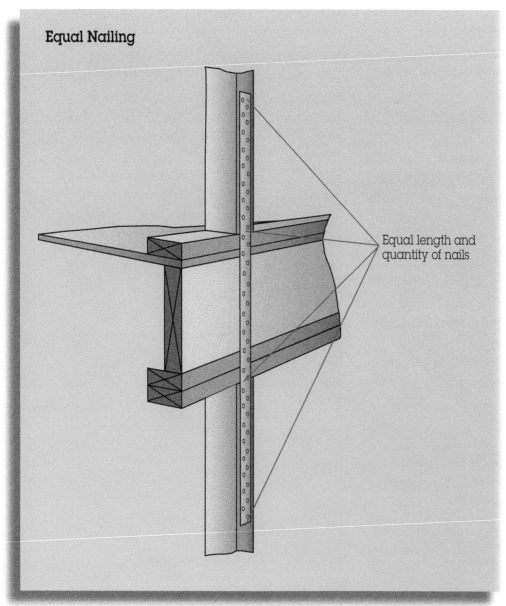

Equal length and quantity of nails

Just as a chain is only as strong as its weakest link, this strap and its ability to hold two walls together is only as good as its weakest side.

Nail Hole Shapes

Round Holes
Purpose: to fasten a connector to wood.
Fill Requirements: always fill, unless noted otherwise.

Obround Holes
Purpose: to make fastening a connector in a tight location easier.
Fill Requirements: always fill.

Hexagonal Holes
Purpose: to fasten a connector to concrete or masonry.
Fill Requirements: always fill when fastening a connector to concrete or masonry.

Diamond Holes
Purpose: to temporarily fasten a connector to make installing it easier.
Fill Requirements: none.

Triangular Holes
Purpose: to increase a connector's strength or to achieve MAX strength.
Fill Requirements: when the Designer specifies max nailing.

Pilot Hole

Pilot Holes
Tooling holes for manufacturing purposes. No fasteners required.

Source: Simpson Strong-Tie Company, Inc.

studs before nailing them into the wall. When the wall sheathing is installed, make sure it is nailed to the hold-down studs using the same nailing pattern that was used for edge nailing. (See "Hold-Down Nailing" illustration.)

Install the hold-downs and bolts, and washers and nuts, as soon as possible. Note, too, that when installing hold-downs after the walls are built, it is more productive to do an entire floor at one time. If the anchor bolts in the concrete do not extend high enough, a coupler nut can be used to extend the length. (See "Coupler Nuts Can Extend Anchor Bolts" illustration.)

As noted previously, the holes drilled for the bolts attaching the hold-down to the studs should not be more than 1/16" bigger than the bolts. However, it is acceptable to oversize the holes you drill for the threaded rod that passes between the floors. This will make installation easier without affecting

strength. (See "Drill Hole Size for Hold-Downs" illustration.)

With all nail-on connection hardware, it is important to use the right size nail. Hardware manufacturer's catalogs indicate nail size appropriate for each piece of hardware. Most catalogs also give some options for nail use.

Hold-downs

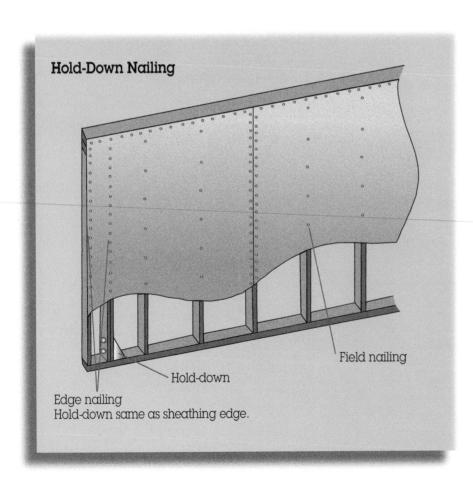

Hold-Down Nailing

Field nailing

Hold-down

Edge nailing
Hold-down same as sheathing edge.

Coupler Nuts Can Extend Anchor Bolts

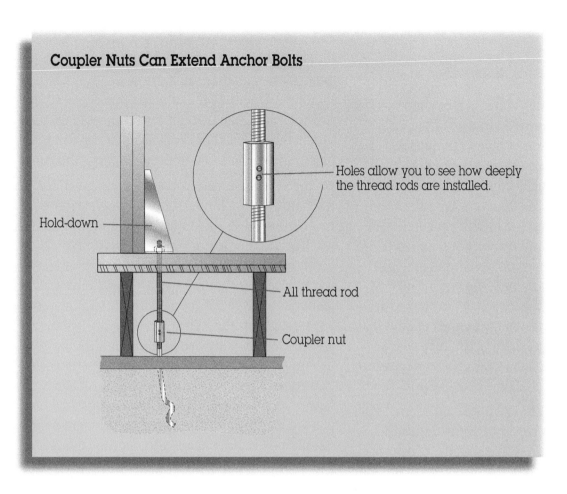

Holes allow you to see how deeply the thread rods are installed.

Hold-down

All thread rod

Coupler nut

Drill Hole Size for Hold-Downs

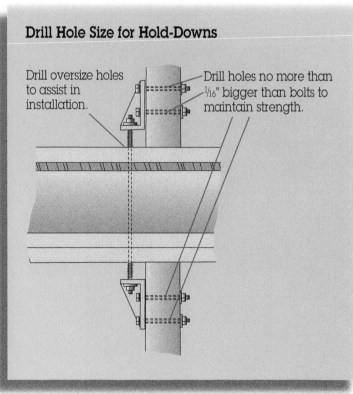

Drill oversize holes to assist in installation.

Drill holes no more than $1/16$" bigger than bolts to maintain strength.

Positive Placement Nail Guns

Earthquakes, hurricanes, and tornados continue to wreak havoc on our wood frame houses and buildings. We will never be able to completely protect against the worst case scenario; however, our codes are continually improving so that we can make our buildings stronger. A big part of this improvement has been the addition of connection hardware. Whereas most connections used to be secured by nails, connections needed to establish shear and diaphragm strength are now secured by hardware. Most of this hardware is fastened with nails and in many cases a large number of nails. For example, where a small framing clip may take 12 nails, a four-foot strap may take 32 nails, depending on the particular size and type of connector.

Because of all the additional hardware nailing, nail gun manufacturers have come out with positive placement nail guns that are specially made for nailing on hardware. There are different styles but they all use the same nails which are different from standard nail guns. The nails are hardened and come in four sizes which are $0.131 \times 1\text{-}\frac{1}{2}"$, $0.148 \times 1\text{-}\frac{1}{2}"$, $0.148 \times 2\text{-}\frac{1}{2}"$, and $0.162 \times 2\text{-}\frac{1}{2}"$. The guns use two methods to find the nail holes in the hardware. One style uses a probe that is placed in the hole, and then the gun directs the nail. In the other style, the nail protrudes so that the nail is placed in the hardware hole before firing the gun.

You need to make sure you use the right nail for the hardware. Each piece of hardware has its own nail requirements. If you use too big a nail you can fracture the steel around the nail hole, and if you use too small a nail you will not develop the appropriate strength needed. Hardware manufacturer specifications note the requirements. For example on the web at strongtie.com, Simpson Strong-Tie Company lists all their hardware with the amount and size of nails needed. There is also a convenient nail replacement chart which lists some nail size substitutions. This is helpful when you are installing hardware that was designed for standard nails but you are using positive placement gun nails. You can find this chart at strongtie.com/products/connectors/nails.asp.

Conclusion

Quality of installation is probably the most important part of framing to withstand the forces of nature. APA (formerly the American Plywood Association, now the Engineered Wood Association) confirmed this fact when it conducted a study of the construction failures in the aftermath of Hurricane Andrew. In the houses they investigated, roofs were the most common failures. Those roof systems most often failed due to lack of proper sheathing nailing.

Wind- and earthquake-resistant framing are important skills for lead framers, and essential to those in susceptible parts of the country. Building codes, along with the designs architects and engineers create to meet code requirements, specify the framing for wind and earthquake resistance. The lead framer must take that information, along with data from connector manufacturers, and ensure that those requirements are met.

Source: APA, The Engineered Wood Products Association

Roof failure as a result of Hurricane Andrew

Chapter Ten
BUILDING CODE REQUIREMENTS

Contents

Chapter Ten

BUILDING CODE REQUIREMENTS

Framers, builders, architects, engineers, and building inspectors alike have contributed to the system of building codes we use today. You should be aware of the codes that apply to the part of the country you are working in, as well as the important features of those codes. This chapter will discuss what you should know about building code requirements.

There are many exceptions to these codes. This book does not cover all exceptions. If you have questions or concerns, it would be good to review the relevant code in the code books, by using the IBC and IRC Framing Index included in this chapter.

Introduction to Building Codes

The Evolution of Building Codes

Although carpentry is one of the oldest professions, framing as we know it today didn't start until 1832 when a man named George Snow wanted to build a warehouse in Chicago. It was difficult to obtain enough large timbers to build the structure using the traditional post and beam method. Being creative (as all good builders and lead framers must be), he cut up the small timbers he had growing on his property into pieces similar to 2 × 4s. He placed them in a repetitive manner, thus creating the first 2 × 4 style walls.

Since then, architects, engineers, builders, building inspectors, and framers have all contributed to the system we use today. Along the way, builders constructed buildings in the way they saw fit. Although this "every man for himself" approach to building gave us structures to live and work in, it did not guarantee that such buildings would last a lifetime, be safe to live and work in, or stand up against earthquakes and hurricanes.

It wasn't until 1915 that a group of building officials decided they needed a standard. That year, the Building Officials and Code Administrators International (BOCA) was established to bring some uniformity to the systems being used.

The IBC

Two other building code agencies appeared not long after: the International Conference of Building Officials (ICBO), and the Southern Building Code Congress International (SBCCI). All three organizations worked to meet the particular needs of their regions of the country.

In the year 2000, these agencies combined their codes to create one common code that would cover the entire country. This code is divided into two books: the *International Residential Code* (IRC), which covers all one- and two-family dwellings and multiple single-family dwellings (townhouses) not more than three stories in height, and the *International Building Code* (IBC), which covers all buildings. Separating the code in this way makes it easier to find the information you need. If you are building only houses, duplexes, or townhouses, you would go straight to the IRC.

There are two ways to comply with the code. The *prescriptive method*, most commonly used, gives specific requirements (such as how many inches on center to space the framing lumber) to build walls that are acceptable. The *performance method* tells us how a person can determine the strength of a wall using properly stamped, graded lumber, and if that strength meets the minimum code requirements.

Because the prescriptive system is most commonly used, it is the one we'll cover here. It applies to conventional construction otherwise known as *platform* or *balloon* framing, which has been developed over the years on job sites, and has been tested and standardized. Prescriptive code requires no "engineering" design by a registered professional, as long as the project is built in compliance with the *International Residential Code* (IRC) or *International Building Code* (IBC).

(Note that with a performance-rated system, you will have a set of plans that you must follow to the letter. These plans come with structural components that must be used exclusively with the plans. Performance-rated codes require design by a registered professional who must specify in accordance with the IRC or IBC.)

A Framer's Code Responsibility

Although it may seem that the codes are written for lawyers instead of framers, framers must be sure that their work complies to code. Note that some areas of the country may not be covered by a statewide, town, city, or county code. (Counties have historically been the jurisdictions controlling code establishment and enforcement.) Note, too, that code-writing organizations are not government agencies, so codes are not enforceable until or unless a government jurisdiction accepts the codes and makes them part of local law.

Code Revisions and Time Delays

Code Revisions

Revisions are important to keep in mind when working with codes. Codes are normally updated annually, and revised versions are published every three years. Typically, the revisions are not major, but it is important to know which code you must comply with. On some jobs the plans will indicate which codes apply. This information can usually be found on the cover page or with the general specifications in the plans. If the applicable code is not shown on the plans, ask the builder, owner, or whoever acquired the building permit about the code.

Time Delays

Another thing to keep in mind is the time that may elapse between when the code-writing organizations publish a revised code and when that code edition becomes the ruling code on the job you are framing. There are delays between when the code agencies certify the new codes and when the local government agencies review and approve them. There can also be delays between the date the permit is issued and the date the job is framed. It is not unusual to be working on plans that are three or four years or more behind the current building code. Although you have to comply with the code that is specified on the plans or that was used when the building permit was approved, you should also understand the current code because, in general, additions to the codes are improvements, or ways that contribute to making a building stronger. After every major earthquake or hurricane, codes have been adjusted and upgraded. By using the latest code, you can feel confident that you are framing with the latest construction knowledge.

Latest Code Used in This Book

This book uses the 2018 edition of the IBC and IRC to explain the major features of codes related to framing. These include structural requirements and life safety issues, and the spreading of fire. Although the code books may seem big and intimidating when you first see them, the number of pages that deal with framing are relatively few.

The following IBC and IRC Framing Index table is a handy list of all the framing sections of either code you might need. It was compiled based on the 2018 code books. In the IRC, the framing information can be found primarily in 4 of the total 43 chapters. In the IBC, 3 of the total 35 chapters deal with framing. The IRC framing chapters are 3, 5, 6, and 8. The IBC chapters containing framing information are 10, 12, and 23.

Important Code Features

What follows are key features of the code, and illustrations presented in a framer-friendly way. If you do a lot of framing, it's a good idea to have a copy of the code book available for reference.

The three major categories used in the IBC are:
- Use and occupancy classification
- Fire-resistance-rated construction classification
- Seismic design categories

In the IBC, the seismic design categories are based on their seismic use group. The categories are A, B, C, D, D[a], E, and F. Although they are similar to the categories in the IRC, there are some differences.

2018 IBC IRC
Framing Index

Framing code	IRC #	IRC page	IBC#	IBC page	Table-Fig.
Floor Framing					
Double joists under bearing partitions	R502.4	141	2308.4.5	517	
Bearing	R502.6	141	2308.4.2.2	509	
Girders			2308.4.1	509	
Minimum lap	R502.6.1	141	2308.4.2.3	509	
Joist support	R502.6.2	141	2308.4.2.3	509	
Lateral support	R502.7	141	2308.4.2.3	509	
Bridging	R502.7.1	141	2308.4.6	517	
Drilling and notching	R502.8	141	2308.4.2.3	509	
Framing around openings	R502.10	141	2308.4.4	509	
Framing around openings - seismic			2308.4.4.1	517	
Wall Framing					
Stud size, height and spacing	R602.3.1	177	2308.5.1	518	IBC-2308.5.1
Stud size, height and spacing					R602.3(5)
Cripple wall stud size	R602.9	190	2308.5.6	520	
Cripple wall bracing			2308.6.6	528	IBC-2308.6.1
Cripple wall seismic			2308.6.8.3	530	
Double and top plate overlap	R602.3.2	177	2308.5.3.2	519	
Drilling and notching	R602.6	178	2308.5.9	520	IRC-602.6(1)&(2)
Drilling and notching			2308.5.10	520	IRC-602.6.1
Headers	R602.7	178	2308.5.5.1	519	602.7(1)(2)(3)
Headers					IBC-2308.4.1.1(1)
Fireblocking	R302.11	62	718.2	155	
Wall bracing	R602.10	190-218			R602.3(1) to
Wall bracing					R602.12.4
Wall bracing			2308.6	520-32	IBC-2308.6.1 to
Wall bracing					IBC-2308.6.7.2(1)
Anchor bolts	R602.11	218			IRC-R602.11.2
Anchor bolts			2308.6.7.3	528	2308.6.8.3
Plate washers	R602.11.1	218	2308.3.1	508	
Rafter Framing					
Ridge board	R802.3	384	2308.7.3	532	
Hips and valleys	R802.4.3	401			
Rafter connections	R802.5.2	401	2308.7.3.1	532	
Rafter bearing	R802.6	407			
Drilling and notching	R802.7	407	2308.7.4	533	
Lateral support	R802.8	407	2308.7.8	534	
Framing around openings	R802.9	408	2308.7.6	533	
Roof tie downs and wind uplift	R802.11	409	2308.7.5	533	
Ceiling Framing					
Ceiling heights	R305.1	64	1207.2	325	
Ceiling framing	R802.5	401	2308.7	532	2308.7.1(1)(2)
Ceiling joist connections	R802.5.2	401	2308.7.3.1	532	
Ceiling joist lapped	R802.5.2.1	407			
Ceiling joists bearing	R802.6	407			
Framing around openings	R802.9	408			

Compiled by the author from the International Residential Building Code Copyright © 2018. The 2018 International Residential Building Code is a copyrighted work of the International Code Council.

Framing code	IRC #	IRC page	IBC#	IBC page	Table-Fig.
Truss Framing					
Truss bracing	R802.10.3	409	2303.4.1.2	486	
Truss alterations	R802.10.4	409	2303.4.5	487	
Attic Access					
Attic access	R807.1	429	1208.2	325	
Stair and Ramp Framing					
Stair landings	R311.7.6	72	1011.6	278	
Stair width	R311.7.1	71	1011.2	276	
Stair treads and risers	R311.7.5	72	1011.5	277	
Stair headroom	R311.7.2	72	1011.3	277	
Spiral stairs	R311.7.10.1	73	1011.1	278	
Curved stairs			1011.9	278	
Handrails	R311.7.8	72	1011.11	279	
Ramps	R311.8	74	1012	280	
Ventilation					
Roof - Attic	R806	427-29	1202.2	321	
Under floor	R408	133	1202.4	322	
Nailing					
Nailing table	R602.3(1)	174-76	2304.1	493	IBC-2304.10.1
					IRC-602.3(1)
Sheathing nailing table	R602.3(1)	176	2304.1	496	IBC-2304.10.1
					IRC-602.3(1)
Prevention of Decay					
Decay map					IRC-R301.2(7)
Pressure treated	R317	79	2304.12.1	498	
Pressure treated joists, girders, and subfloor			2304.12.1.1	499	
Pressure treated framing			2304.12.1.3	499	
Pressure treated sleepers and sills			2304.12.1.4	499	
Girder ends at masonry			2304.12.2.1	499	
Pressure treated post and columns	R317.1.4	79	2304.12.2.2	499	
Pressure treated post and columns			2304.12.3.1	499	
Pressure treated laminated timbers			2304.12.2.4	499	
Pressure treated wood contact with ground			2304.12.3	499	
Pressure treated wood structural members	R317.1.5	79	2304.12.2.3	499	
Pressure treated wood structural members			2304.12.2.5	499	
Termite protection					
Termite protection	R318	80	2304.12.4	500	
Termite probability map	R301.2(7)	50			
Miscellaneous					
Wind limitations design map	R301.2(5)A&B	46-47			
Seismic design map	R301.2(2)(3)	35-44			
Minimum fixture clearance bath and shower	R307	65			IRC-307.1
Framing around flues and chimneys			2304.5	487	
SIPs (Structural Insulated Panel) Walls	R610	350-51			
Safety					
Safety			33	567	

Compiled by the author from the International Residential Building Code Copyright © 2018. The 2018 International Residential Building Code is a copyrighted work of the International Code Council.

Framing According to Code

Floor Framing

Following are the code requirements and instructions related to floor framing:

- Double joists are required under parallel bearing walls.

- If pipes penetrate floors where double joists are required, the joists must be separated and have full-depth, solid blocks not less than 2" at least every 4' along their length.

- Bearing for joists must be 1-½" minimum on wood or steel, and 3" minimum on concrete or masonry.

- Where joists lap, there must be a minimum lap of 3" or a wood or metal splice of equal strength.

- The ends of joists must be kept from turning by using 2" nominal full-depth solid blocking or by attaching them to a full depth header, band, rim joist, or adjoining stud.

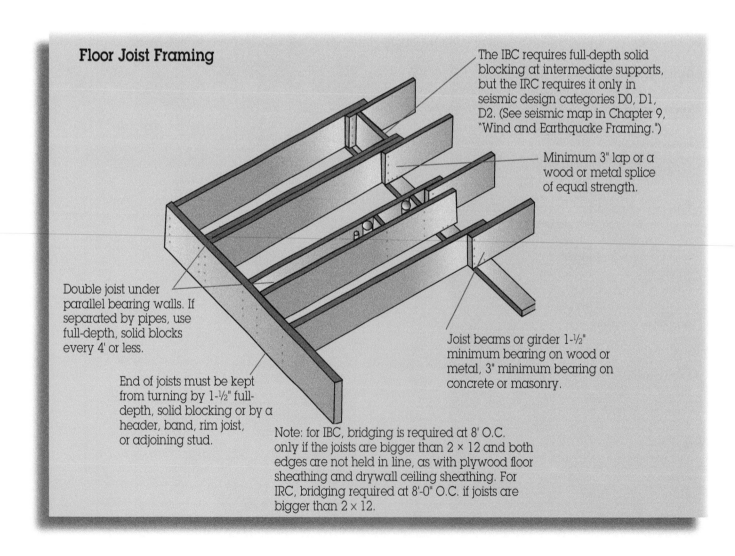

Floor Joist Framing

The IBC requires full-depth solid blocking at intermediate supports, but the IRC requires it only in seismic design categories D0, D1, D2. (See seismic map in Chapter 9, "Wind and Earthquake Framing.")

Minimum 3" lap or a wood or metal splice of equal strength.

Double joist under parallel bearing walls. If separated by pipes, use full-depth, solid blocks every 4' or less.

End of joists must be kept from turning by 1-½" full-depth, solid blocking or by a header, band, rim joist, or adjoining stud.

Joist beams or girder 1-½" minimum bearing on wood or metal, 3" minimum bearing on concrete or masonry.

Note: for IBC, bridging is required at 8' O.C. only if the joists are bigger than 2 × 12 and both edges are not held in line, as with plywood floor sheathing and drywall ceiling sheathing. For IRC, bridging required at 8'-0" O.C. if joists are bigger than 2 × 12.

- Full-depth, solid blocking is required at intermediate supports in IRC seismic design categories D_0, D_1, and D_2. (See seismic maps in Chapter 9.)

- Bridging at 8' O.C. is required only when joists are larger than 2 × 12. IBC does not require bridging if held in line, as with plywood floor sheathing and drywall sheathing.

The "Floor Joists—Anchor or Ledger" illustration below shows how joists framing into girders must be supported by framing anchors or a 2 × 2 or larger ledger.

Engineered wood products, such as I-joists, can be notched according to the manufacturer's specifications. (See Chapter 8 for more on engineered wood products.)

The "Framing Floor Openings" illustration shows the following code requirements and instructions related to framing around openings in floors.

- If the header joists are more than 4', the header joists and trimmer joists should be doubled.

- If the distance from the bearing point of a trimmer joist to the header joist is more than 3', the trimmer joists should be doubled.

- If the header joist is greater than 6', hangers must be used.

- If the tail joists are more than 12', use framing anchors or a 2 × 2 ledger.

The "Seismic Floor Opening Framing" illustration shows what to do if you are building in IBC seismic design categories B, C, D, or E and the opening is greater than 4' perpendicular to the joists. In such cases, you must provide blocking beyond the headers, and metal ties must be used to connect the headers with the blocks.

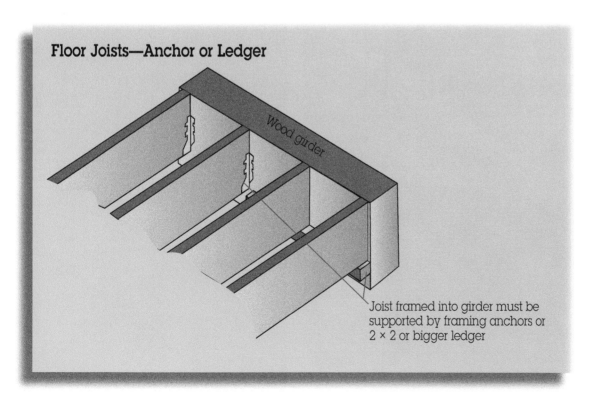

Floor Joists—Anchor or Ledger

Wood girder

Joist framed into girder must be supported by framing anchors or 2 × 2 or bigger ledger

2 × Floor Joists—Drilling and Notching Requirements

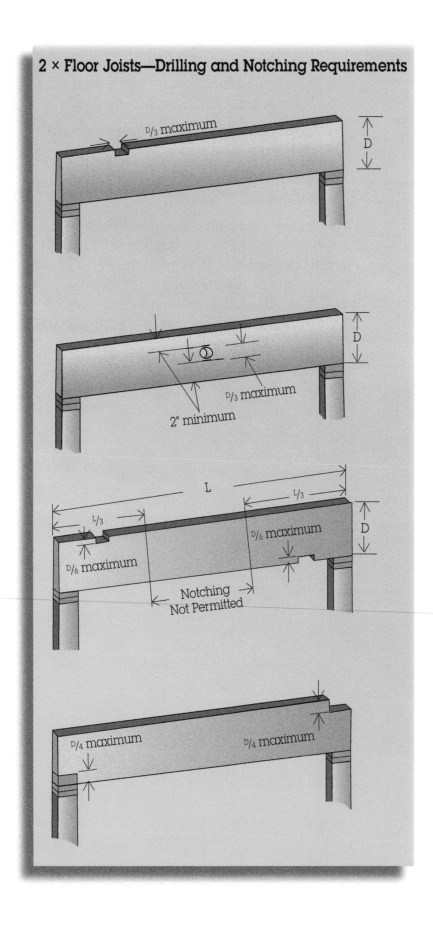

Framing Floor Openings

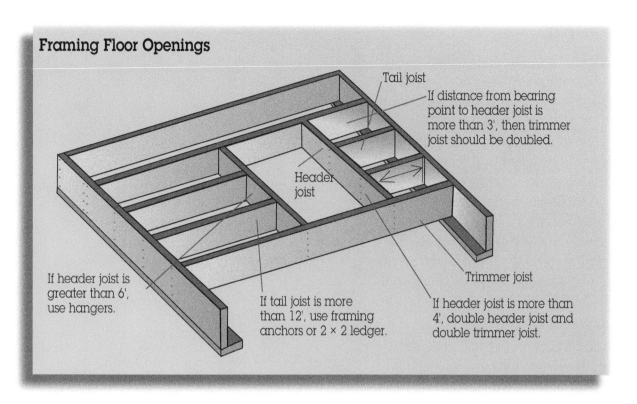

Tail joist

If distance from bearing point to header joist is more than 3', then trimmer joist should be doubled.

Header joist

Trimmer joist

If header joist is greater than 6', use hangers.

If tail joist is more than 12', use framing anchors or 2 × 2 ledger.

If header joist is more than 4', double header joist and double trimmer joist.

Seismic Floor Opening Framing with an Opening Greater than 4' Perpendicular to the Joists in Seismic Design Categories B, C, D and E

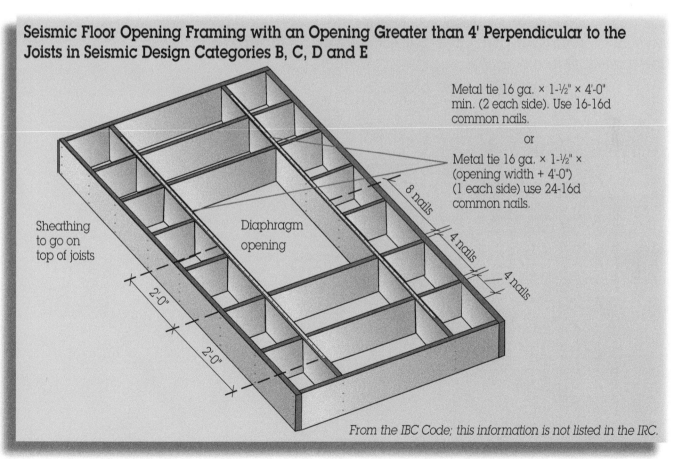

Metal tie 16 ga. × 1-½" × 4'-0" min. (2 each side). Use 16-16d common nails.

or

Metal tie 16 ga. × 1-½" × (opening width + 4'-0") (1 each side) use 24-16d common nails.

8 nails

4 nails

4 nails

Sheathing to go on top of joists

Diaphragm opening

2'-0"

2'-0"

From the IBC Code; this information is not listed in the IRC.

Wall Framing

Stud spacing should be shown on the plans, but it is still good to be familiar with the code limitations. For 2 × 4 studs less than 10 feet tall, the maximum stud spacing is 24" O.C., provided the wall is supporting one floor or a roof and ceiling only. For the support of one floor, a roof, and ceiling, 16" O.C. is the maximum. To support two floors, a roof, and a ceiling with a maximum spacing of 16" O.C. and height of 10', a minimum of 3 × 4 studs must be used. If studs are 2 × 6, a wall can support one floor, a roof, and ceiling at 24" O.C., or two floors, a roof, and ceiling at 16" O.C. Again, this stud spacing only applies to walls that don't exceed 10' in height. (See "Stud—Spacing and Size" illustration.)

Cripple walls less than 4' in height should be framed with studs at least as big as those used in the walls above them. If the cripple walls are higher than 4', then the studs need to be at least the size required for supporting an additional floor level (as described in previous paragraph). (See "Foundation Cripple Walls" illustration.)

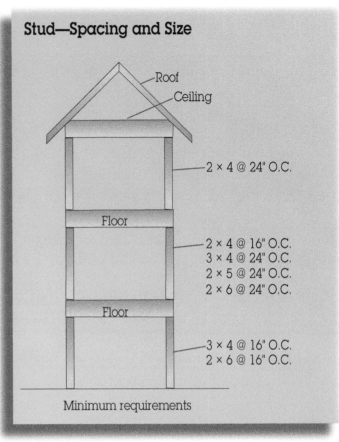

Stud—Spacing and Size

Roof
Ceiling
2 × 4 @ 24" O.C.
Floor
2 × 4 @ 16" O.C.
3 × 4 @ 24" O.C.
2 × 5 @ 24" O.C.
2 × 6 @ 24" O.C.
Floor
3 × 4 @ 16" O.C.
2 × 6 @ 16" O.C.

Minimum requirements

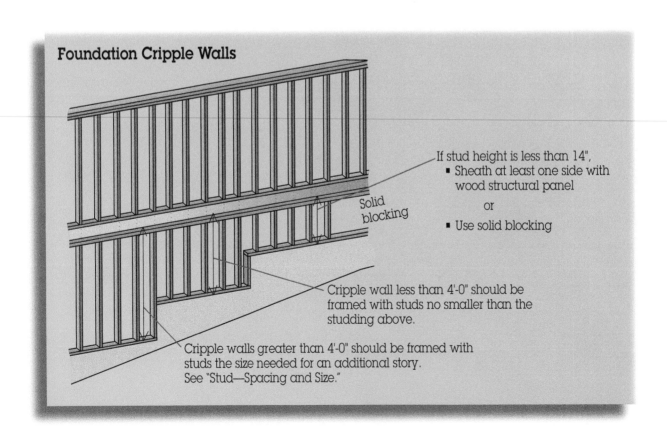

Foundation Cripple Walls

Solid blocking

If stud height is less than 14",
- Sheath at least one side with wood structural panel

or

- Use solid blocking

Cripple wall less than 4'-0" should be framed with studs no smaller than the studding above.

Cripple walls greater than 4'-0" should be framed with studs the size needed for an additional story. See "Stud—Spacing and Size."

216

Double plates are needed on top plates for bearing and exterior walls. The end joints of the top plates and double plates should be offset by at least 48". The IRC allows a 24" offset at nonstructural interior walls. The end joints need to be nailed with at least eight 16d nails or twelve 3" × 0.131" nails on each side of the joint. A single top plate may be used if the plates are tied together at the joints, intersecting walls, and corners with 3" × 6" galvanized steel plates or the equivalent, and all rafters, joists, or trusses are centered over the studs. (See "Walls, Top and Double Plate" illustration.)

Allowable **drilling** and **notching** is different for bearing or exterior walls, and for *non-bearing* or interior walls. Bearing or exterior walls can be notched up to 25% of the width of the stud and drilled up to 40% of the stud provided that the hole is at least ⅝" away from the edge. With interior *non-bearing* walls, the percentages are 40% for notches and 60% for drilling. (See "Drilling and Notching Studs, Exterior and Bearing Walls" and "Drilling and Notching Studs, Interior Nonbearing Walls" illustrations later in this chapter.)

Header sizes for exterior and bearing walls should be specified on the plans. For *nonbearing* walls, a flat 2 × 4 may be used as a header for a maximum of up to 8' span where the height above the header to the top plate is 24" or less. (See "Header for Nonbearing Walls" illustration later in this chapter.)

Fireblocking refers to material you install to prevent flames from traveling through concealed spaces between areas of a building. The location of fireblocks is sometimes difficult to understand. It helps to think of where flames would be able to go. A 1-½"-thick piece of wood can create a fireblock. If you place a row of

these blocks in a wall, you create a deterrent for the vertical spread of fire. Vertical and horizontal fireblocks are required in walls at least every 10'. (See "Fireblocking Vertical" and "Fireblocking Horizontal" illustrations later in this chapter.)

In a "party wall" construction, where you have two walls next to each other, you can create a fireblock by installing a stud in the space between the studs in the two adjoining walls. This creates a vertical fireblock. Note that ½" gypsum board can also be used to create this type of fireblock.

Fireblocking is required between walls, floors, ceilings, and roofs. Typically, the drywall covering creates this fireblock. If it doesn't, then fireblocking is needed. Where fireblocking is required behind the ledger, it can be installed at the interconnections of any concealed vertical and horizontal space like that which occurs at soffits, drop ceilings, or cove ceilings. (See "Fireblocking at Interconnections" illustration later in this chapter.)

Stair stringers must be fireblocked at the top and bottom of each run and between studs along the stair stringers if the walls below the stairs are unfinished.

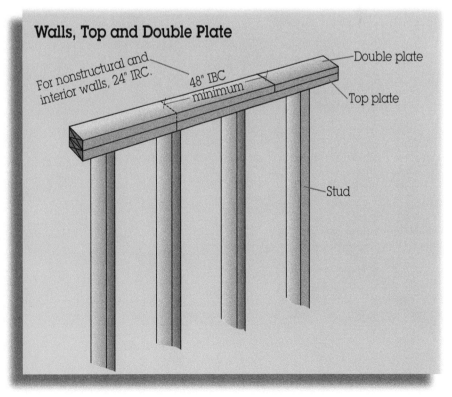

Walls, Top and Double Plate

For nonstructural and interior walls, 24" IRC.

48" IBC minimum

Double plate

Top plate

Stud

217

Drilling and Notching Studs, Exterior and Bearing Walls

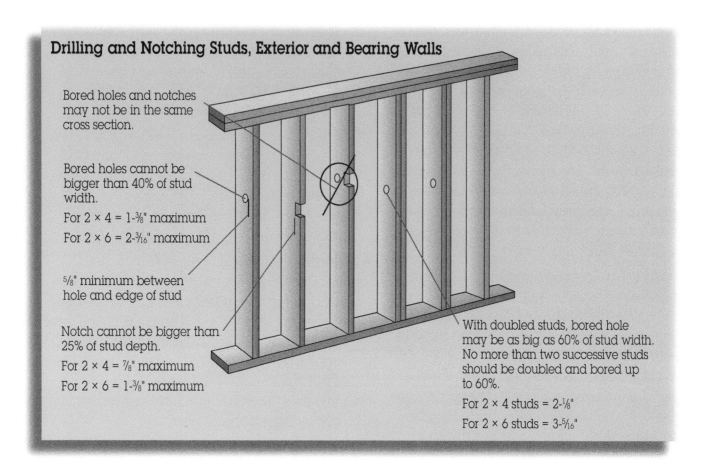

Bored holes and notches may not be in the same cross section.

Bored holes cannot be bigger than 40% of stud width.

For 2 × 4 = 1-⅜" maximum

For 2 × 6 = 2-³⁄₁₆" maximum

⅝" minimum between hole and edge of stud

Notch cannot be bigger than 25% of stud depth.

For 2 × 4 = ⅞" maximum

For 2 × 6 = 1-⅜" maximum

With doubled studs, bored hole may be as big as 60% of stud width. No more than two successive studs should be doubled and bored up to 60%.

For 2 × 4 studs = 2-⅛"

For 2 × 6 studs = 3-⁵⁄₁₆"

Drilling and Notching Studs, Interior Nonbearing Walls

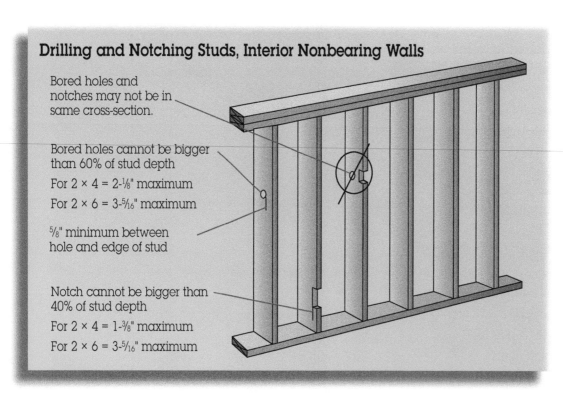

Bored holes and notches may not be in same cross-section.

Bored holes cannot be bigger than 60% of stud depth

For 2 × 4 = 2-⅛" maximum

For 2 × 6 = 3-⁵⁄₁₆" maximum

⅝" minimum between hole and edge of stud

Notch cannot be bigger than 40% of stud depth

For 2 × 4 = 1-⅜" maximum

For 2 × 6 = 3-⁵⁄₁₆" maximum

Header for Nonbearing Walls

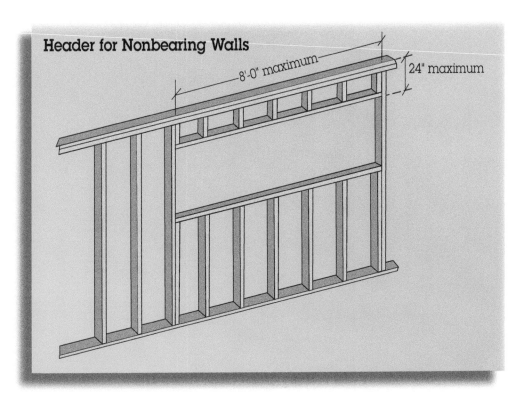

8'-0" maximum

24" maximum

Fireblocking Vertical

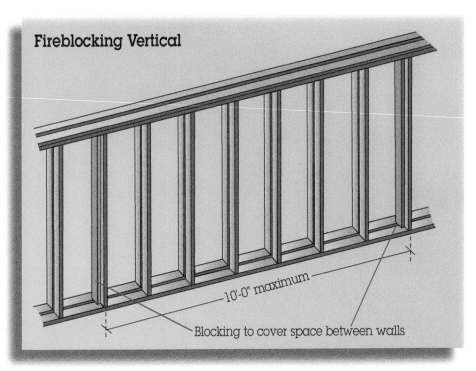

10'-0" maximum

Blocking to cover space between walls

Fireblocking Horizontal

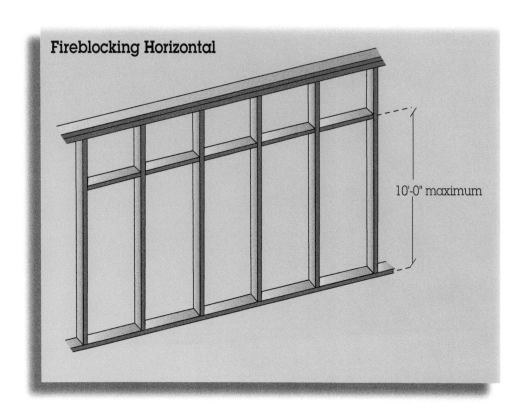

10'-0" maximum

Fireblocking at Interconnections

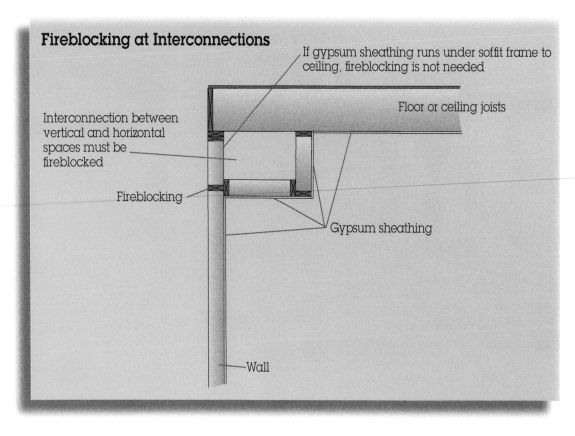

If gypsum sheathing runs under soffit frame to ceiling, fireblocking is not needed

Floor or ceiling joists

Interconnection between vertical and horizontal spaces must be fireblocked

Fireblocking

Gypsum sheathing

Wall

Wall Bracing

Wall bracing is needed to keep buildings from falling. Sheathing the exterior walls is a typical way to provide bracing. The architect, engineer, or whoever creates the plans will specify when any special bracing is needed. Although you don't need to know everything about wall bracing, it is good to have a basic understanding of it.

Two common exceptions to these methods are: (1) the short wall often used for garages, and (2) the 24" wide corner wall. Note that cripple walls have their own requirements.

The IBC states that braced wall panels must be clearly indicated on the plans. However, this is not always the case in the real world. Although shear walls are usually marked on the plans, braced wall panels often are not.

The IBC and IRC contain a table that shows braced wall panel limitations and requirements. The limitations are related to the seismic design category, and to how many stories are built on top of the walls.

Where braced wall lines rest on concrete or masonry foundations, they must have **anchor bolts** that are not less than ½" in diameter or a code-approved anchor strap. The anchor bolts or straps should be spaced not more than 6' apart (or not more than 4' apart if the building is over two stories).

Each piece of wall plate must contain at least two bolts or straps. There must be one between 4" and 12" from each end of each piece and placed in the middle third of the plate. A nut and washer must be tightened on each bolt. IBC seismic design categories D and E require 0.229" × 3" × 3" plate washers.

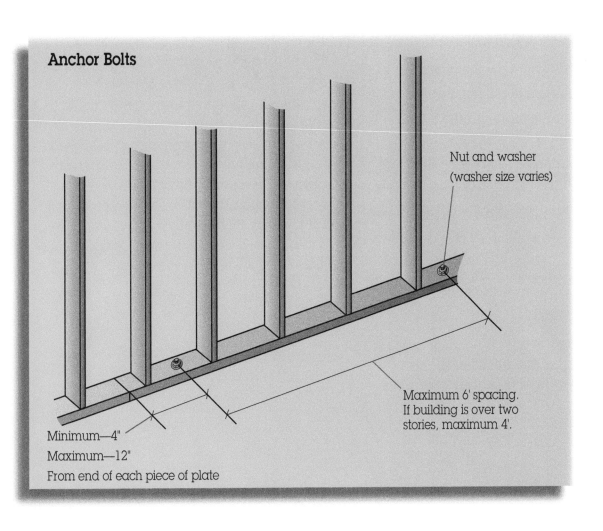

Anchor Bolts

Nut and washer (washer size varies)

Maximum 6' spacing. If building is over two stories, maximum 4'.

Minimum—4"
Maximum—12"
From end of each piece of plate

In IRC seismic design categories D_0, D_1, D_2, and townhouses in category C, braced walls require 0.229" × 3" × 3" plate washers. (See "Anchor Bolts" illustration.)

The 16 basic construction methods for braced wall panels are as follows:

1. LIB—Let-in-bracing
2. DWB—Diagonal wood boards
3. WSP—Wood structural panel
4. BV-WSP—Wood structural panel or masonry veneer with stone
5. SFB—Structural fiberboard sheathing
6. GB—Gypsum board
7. PBS—Particleboard sheathing
8. PCP—Portland cement plaster
9. HPS—Hardboard panel siding
10. ABW—Alternate braced wall
11. PFH—Intermittent portal frame
12. PFG—Intermittent portal frame at garage
13. CS-WSP—Continuously sheathed wood structural panel
14. CS-G—Continuously sheathed wood structural panel adjacent to garage opening
15. CS-PF—Continuously sheathed portal frame
16. CS-SFB—Continuously sheathed structural fiberboard

Rafter Framing

Ridge boards must be at least 1" nominal in width and must be as deep as the cut end of the rafter. Hip and valley rafters must be at least 2" nominal

and must be as deep as the cut ends of the rafters connecting to the hip or the valley. **Gusset plates** as a tie between rafters may be used to replace a ridge board.

Rafters must have a bearing surface similar to that of joists at their end supports. Bearing needs to be 1-½" on wood or metal and not less than 3" on masonry or concrete.

Drilling and notching have the same limitations for rafters as they do for floor joists. (See "Rafter Drilling and Notching" illustration.)

To prevent rotation of rafter framing members, **lateral support** or **blocking** must be provided for rafters and ceiling joists larger than 2 × 10s.

Bridging must be provided for roofs or ceilings larger than 2 × 12. The bridging may be solid blocking, diagonal bridging, or a continuous 1" × 3" wood strip nailed across the ceiling joists or rafters at intervals not greater than 8'. Bridging is not needed if the ceiling joists or rafters are held in line for the entire length with, for example, sheathing on one side and gypsum board on the other.

When rafters are used to frame the roof, the walls that the rafters bear on must be tied together by a connection to keep them from being pushed out. If these walls are not tied together, then the ridge board must be supported by or framed as a beam in order to support the ridge. Ceiling joists are typically used to tie the walls together. The ceiling joists must be tied to the rafters, the walls, and any lapping ceiling joists. (See ceiling joists.)

Rafter Drilling and Notching

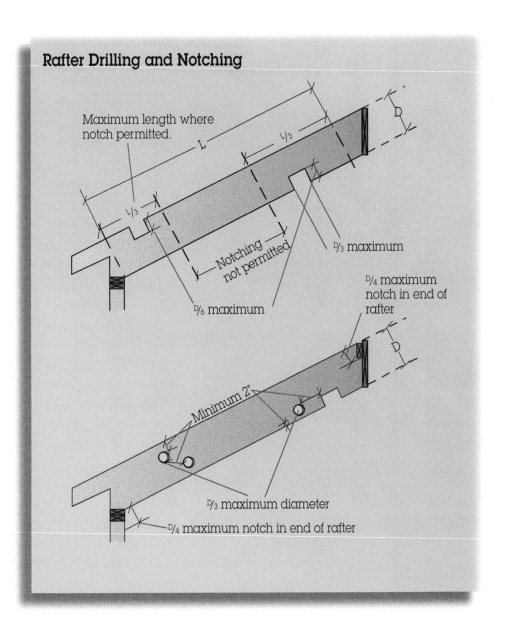

Maximum length where notch permitted.

L

L/3

L/3

D

D/3 maximum

Notching not permitted

D/6 maximum

D/4 maximum notch in end of rafter

D

Minimum 2"

D/3 maximum diameter

D/4 maximum notch in end of rafter

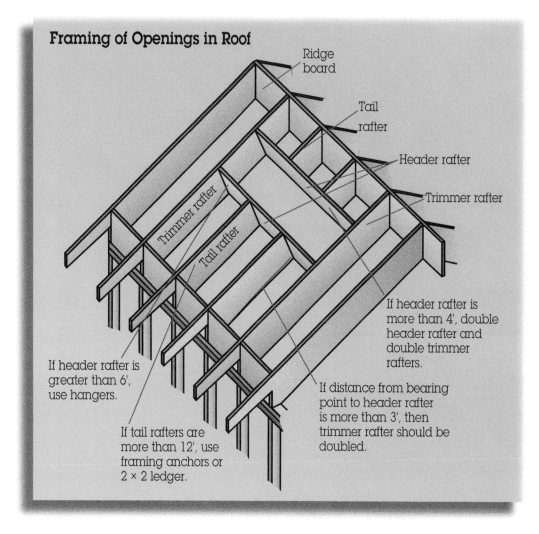

Framing of Openings in Roof

Ridge board

Tail rafter

Header rafter

Trimmer rafter

Trimmer rafter

Tail rafter

If header rafter is more than 4', double header rafter and double trimmer rafters.

If header rafter is greater than 6', use hangers.

If distance from bearing point to header rafter is more than 3', then trimmer rafter should be doubled.

If tail rafters are more than 12', use framing anchors or 2 × 2 ledger.

Ceiling Framing

Ceiling joists must have **bearing support** similar to that of rafters. The bearing must be 1-½" on wood or metal, and not less than 3" on masonry or concrete.

The most important thing to remember about ceiling joists is that if they are used to tie the rafter-bearing walls at opposite ends of the building, then those joists must be **securely attached** to the walls, to the rafters, and to each other at the laps. If the ceiling joists do not run parallel with the rafters, an equivalent rafter tie must be installed to provide a continuous tie across the building.

The IRC calls for a **minimum ceiling clearance** of 7'. The IBC requires 7'-6" with the exception of bathrooms, kitchens, laundry, and storage rooms, where it can be 6'-8" IRC, 7'-0" IBC.

There are three exceptions to this rule. First, in the IBC one- and two-family dwellings, beams or girders can project 6" below the required ceiling height if they are spaced more than 4' apart. In the IRC, they may project to 6'-4" from the finished floor. The second exception is in the IRC for basements without habitable spaces. These may have a minimum height of 6'-8" and may have beams, girders, ducts, and other obstructions at 6'-4" in height. The third exception is for a sloped ceiling. Fifty percent of the sloped ceiling room area can be less than the minimum ceiling height. However, any portion of the room less than 5' in height cannot be included in figuring the room area. (See "Ceiling Heights" illustration later in this chapter.)

Truss Framing

Trusses are an engineered product. This means that an engineer or design professional must design them for each job to form a roof/ceiling system. Components and members of the trusses **should not be notched, cut, drilled, spliced,** or **altered** in any way without the approval of a registered design professional.

Ceiling Joists

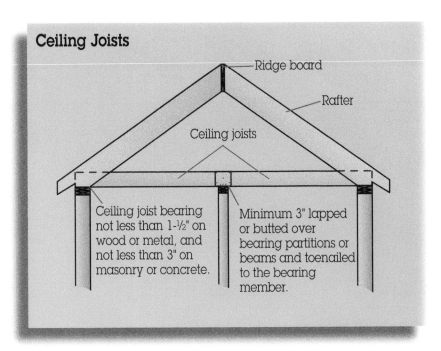

Ridge board

Rafter

Ceiling joists

Ceiling joist bearing not less than 1-½" on wood or metal, and not less than 3" on masonry or concrete.

Minimum 3" lapped or butted over bearing partitions or beams and toenailed to the bearing member.

Attic Access & Ceiling Heights

An attic access must be provided if the attic area exceeds 30 square feet, and the height is at least 30". This opening must be at least 22" × 30", and there must be a height of at least 30" at the access opening. (See "Attic Access" illustration below.)

Attic Access

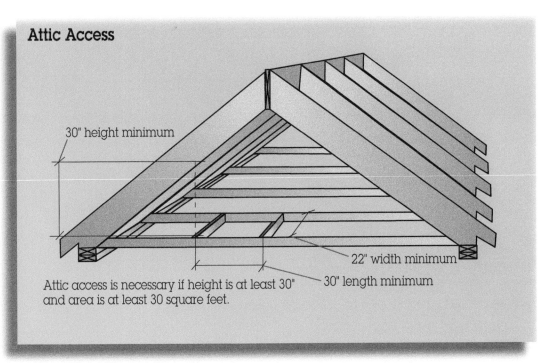

30" height minimum

22" width minimum

30" length minimum

Attic access is necessary if height is at least 30" and area is at least 30 square feet.

Ceiling Heights

Habitable rooms, hallways, corridors, bathrooms, toilet rooms, laundry rooms and basements

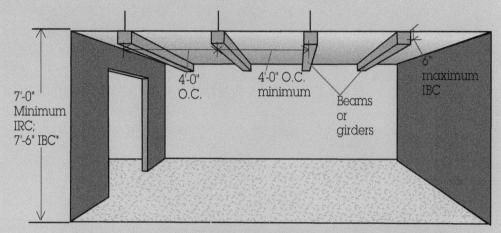

7'-0"
Minimum
IRC;
7'-6" IBC*

4'-0"
O.C.

4'-0" O.C.
minimum

Beams
or
girders

6"
maximum
IBC

Basements without habitable spaces

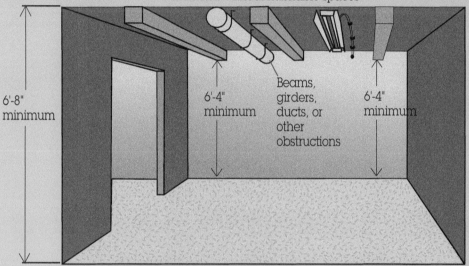

6'-8"
minimum

6'-4"
minimum

Beams,
girders,
ducts, or
other
obstructions

6'-4"
minimum

Sloped ceiling

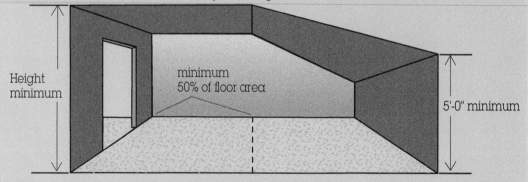

Height
minimum

minimum
50% of floor area

5'-0" minimum

*7'-6" required by IBC for occupiable spaces, habitable spaces, and corridor.

7'-0" required for bathrooms, toilet rooms, kitchens, storage rooms, and laundry rooms.

Stair and Ramp Framing

The **width** of stairs must be a minimum of 36" from finish to finish. Handrails may project into the 36" a maximum of 4-½" on each side. (See "Stairs" illustration.)

Two sets of **tread** and **riser dimensions** apply to minimum and maximum requirement. One set is for Group R-3, Group R-2, and Group 4 (houses, apartments, dormitories, non-transient housing). The other is for all other groups. The first set requires a maximum riser height of 7-¾" and a minimum tread depth of 10", while the second requires a maximum riser height of 7", a minimum riser height of 4", and a minimum tread depth of 11".

The **variation** in riser height within any flight of stairs must not be more than ⅜" from finish tread to finish tread. The variation in tread depth within any flight of stairs cannot be more than ⅜" from the finish riser to the nose of the tread.

Headroom for stairways must have a minimum finish clearance of 6'-8", measured vertically from a line connecting the edge of the nosings.

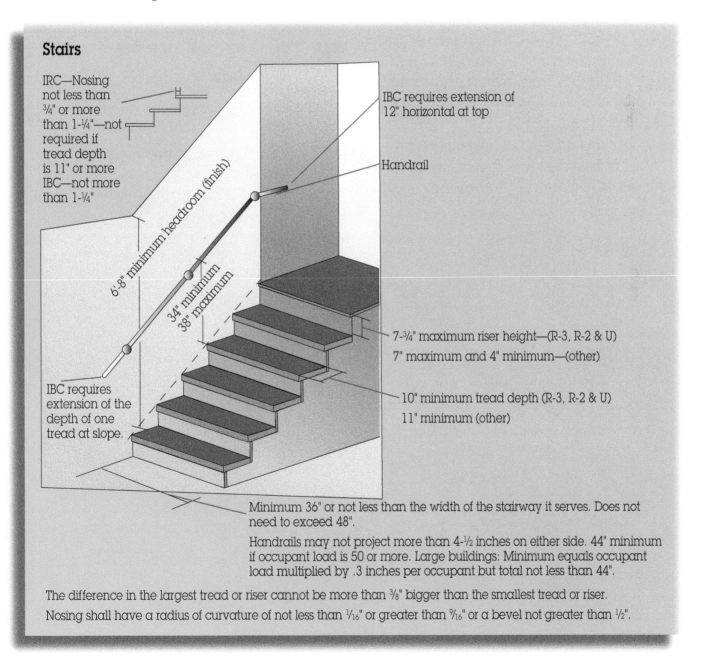

Stairs

IRC—Nosing not less than ¾" or more than 1-¼"—not required if tread depth is 11" or more IBC—not more than 1-¼"

6'-8" minimum headroom (finish)

34" minimum 38" maximum

IBC requires extension of 12" horizontal at top

Handrail

IBC requires extension of the depth of one tread at slope.

7-¾" maximum riser height—(R-3, R-2 & U)
7" maximum and 4" minimum—(other)

10" minimum tread depth (R-3, R-2 & U)
11" minimum (other)

Minimum 36" or not less than the width of the stairway it serves. Does not need to exceed 48".

Handrails may not project more than 4-½ inches on either side. 44" minimum if occupant load is 50 or more. Large buildings: Minimum equals occupant load multiplied by .3 inches per occupant but total not less than 44".

The difference in the largest tread or riser cannot be more than ⅜" bigger than the smallest tread or riser.

Nosing shall have a radius of curvature of not less than 1/16" or greater than 9/16" or a bevel not greater than ½".

Handrails for stairs must have a height of no less than 34" and no more than 38", measured vertically from a line created by joining the nosing on the treads.

Stairway **landings** must be provided for each stairway at the top and bottom. The width each way of the landing must not be less than the width of the stairway it serves. The landing's minimum dimension in the direction of travel cannot be less than 36", but does not need to be greater than 48" for a stair having a straight run. (See "Stair Landing" illustration.)

Curved stairways should have a minimum tread depth at a point 12" from the edge of the tread at its narrowest point of not less than 10" for the IRC and 11" for the IBC. The minimum depth at any point must be 6" for IRC and 10" IBC. (See "Curved Stairs" illustration.)

Spiral stairways must have a minimum width of 26". Each tread must have a minimum tread width of 6-¾" at a point 12" from the narrow edge of the tread. The rise must be no more than 9-½". All treads must be identical. The headroom is a minimum of 6'-6". (See "Spiral Stairs" illustration.)

The maximum slope on a **ramp** is 8%, or one unit of rise for 12 units of run. Some exceptions (where technically infeasible) are available for slope of 1 unit vertical in 8 units horizontal—12-½%. Handrails must be provided when the slope exceeds 8.33%, or one unit of rise and 12 units of run.

The **minimum headroom** on any part of a ramp is 6'-8".

A minimum 36" × 36" **landing** is required at the top and bottom of a ramp and where there is any door, or where the ramp changes direction. The actual minimum landing dimensions will depend on the building use and occupant capacity. This minimum does not apply to non-accessible housing.

The maximum **total rise** of any ramp cannot be more than 30" between level landings. (See "Ramps" illustration later in this chapter.)

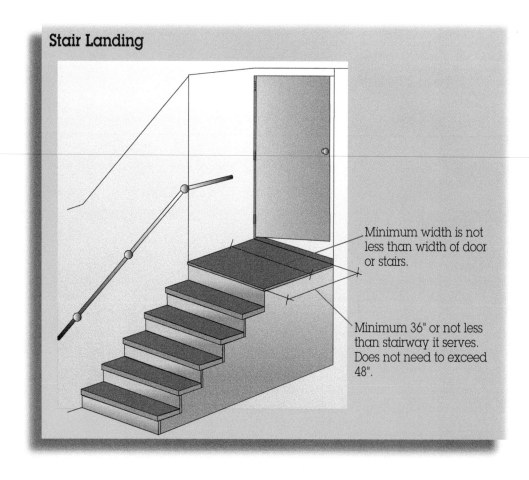

Stair Landing

Minimum width is not less than width of door or stairs.

Minimum 36" or not less than stairway it serves. Does not need to exceed 48".

Curved and Spiral Stairs

Curved Stairs

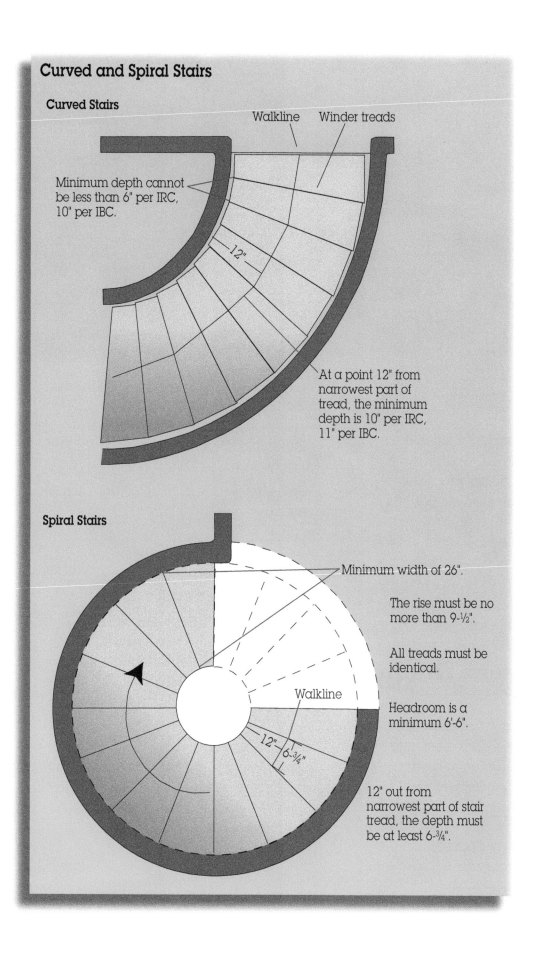

Walkline Winder treads

Minimum depth cannot be less than 6" per IRC, 10" per IBC.

12"

At a point 12" from narrowest part of tread, the minimum depth is 10" per IRC, 11" per IBC.

Spiral Stairs

Minimum width of 26".

The rise must be no more than 9-½".

All treads must be identical.

Headroom is a minimum 6'-6".

Walkline

12" 6-¾"

12" out from narrowest part of stair tread, the depth must be at least 6-¾".

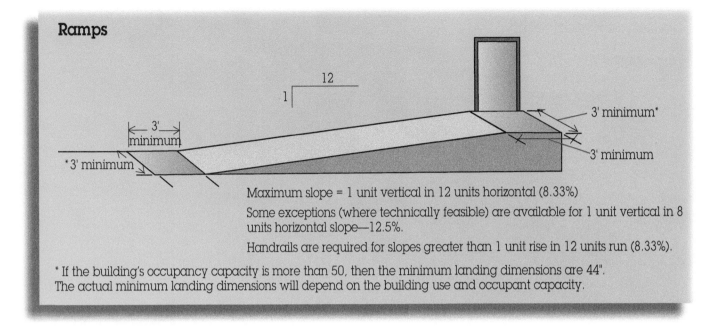

Ramps

12
1

3' minimum

*3' minimum

3' minimum*

3' minimum

Maximum slope = 1 unit vertical in 12 units horizontal (8.33%)

Some exceptions (where technically feasible) are available for 1 unit vertical in 8 units horizontal slope—12.5%.

Handrails are required for slopes greater than 1 unit rise in 12 units run (8.33%).

* If the building's occupancy capacity is more than 50, then the minimum landing dimensions are 44". The actual minimum landing dimensions will depend on the building use and occupant capacity.

Ventilation

Ventilation is required so that condensation does not occur on the structural wood, causing dry rot and the deterioration of the building. Cross ventilation is required in crawl spaces, attics, and in enclosed rafter spaces. In rafter spaces between the insulation and the roof sheathing, there must be at least 1" clear space.

The **total area** of the space to be ventilated cannot be more than 150 times the size of the area of the venting. (Both are measured in square feet.)

Protection from Decay

Moisture and warm air are catalysts of fungus, which causes dry rot that can destroy a building. In addition to calling for ventilation to control moisture, the code also requires decay-resistant wood wherever moisture can come in contact with structural wood. Some areas of the country are more conducive to decay than others. The code requires naturally durable wood or preservative-treated wood in the following situations:

- Wood joist or the bottom of the wood floor structure if less than 18" from exposed ground. (See "Joists and Girder Protection" illustration.)

- Wood girders if closer than 12" from exposed ground.

- Wall plates, mudsills, or sheathing that rests on concrete or masonry exterior walls less than 8" from exposed ground. (See "Exterior Wall Decay Protection" illustration.)

- Sills or sleepers that rest on a concrete or masonry slab in direct contact with the ground, unless separated from the slab by an impervious moisture barrier. (See "Decay Protection from Slab" illustration later in this chapter.)

- The ends of wood girders entering exterior masonry or concrete walls having less than ½" clearance on tops, sides, and ends. (See "Ends of Girders in Masonry or Concrete" illustration.)

- Wood furring strips or framing members attached directly to the interior of exterior concrete or masonry walls below grade.

- Wood siding less than 6" from exposed ground or less than 2" measured vertically from concrete steps, porch slabs, patio slabs, and similar horizontal surfaces exposed to the weather.

- Posts or columns that support permanent structures and are themselves supported by a masonry concrete slab or footing in direct contact with the ground. (See "Post and Column Decay-Resistant Wood" illustration.)

Joist and Girder Protection

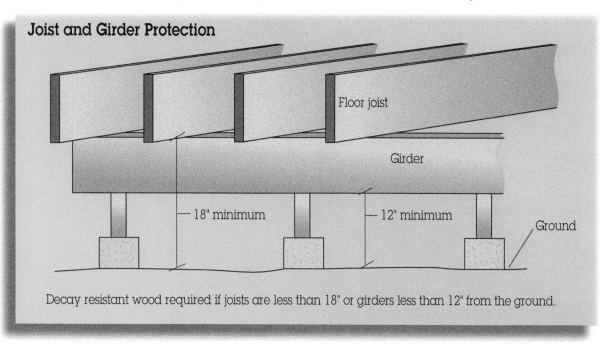

Floor joist

Girder

18" minimum

12" minimum

Ground

Decay resistant wood required if joists are less than 18" or girders less than 12" from the ground.

Exterior Wall Decay Protection

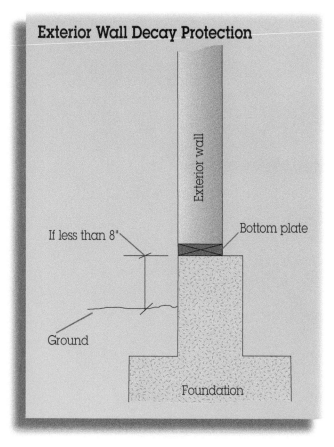

Exterior wall

Bottom plate

If less than 8"

Ground

Foundation

Ends of Girders in Masonry or Concrete

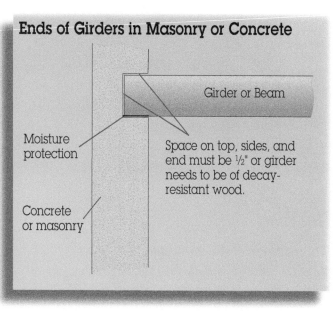

Girder or Beam

Moisture protection

Space on top, sides, and end must be ½" or girder needs to be of decay-resistant wood.

Concrete or masonry

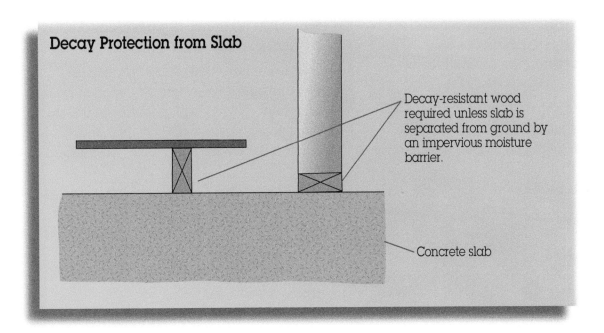

Decay Protection from Slab

Decay-resistant wood required unless slab is separated from ground by an impervious moisture barrier.

Concrete slab

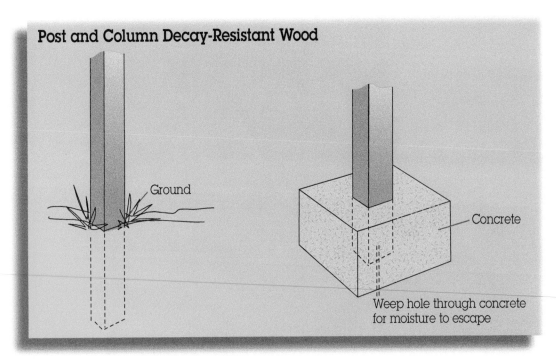

Post and Column Decay-Resistant Wood

Ground

Concrete

Weep hole through concrete for moisture to escape

Nailing

Nailing is one of the most important parts of framing. Table 2304.10.1, Fastening Schedule (see IBC Nailing Table), is taken directly from the IBC 2018. The table shows use of alternate nails. The 3" × 0.131" nail is the most common nail gun nail used for framing.

TABLE 2304.10.1
FASTENING SCHEDULE

DESCRIPTION OF BUILDING ELEMENTS	NUMBER AND TYPE OF FASTENER	SPACING AND LOCATION
Roof		
1. Blocking between ceiling joists, rafters or trusses to top plate or other framing below	3-8d common (2-½" × 0.131"); or 3-10d box (3" × 0.128"); or 3-3" × 0.131" nails; or 3-3" 14 gage staples, ⁷⁄₁₆" crown	Each end, toenail
Blocking between rafters or truss not at the wall top plate, to rafter or truss	2-8d common (2-½" × 0.131") 2-3" × 0.131" nails 2-3" 14 gage staples	Each end, toenail
	2-16 d common (3-½" × 0.162") 3-3" × 0.131" nails 3-3" 14 gage staples	End nail
Flat blocking to truss and web filler	16d common (3-½" × 0.162") @ 6" o.c. 3" × 0.131" nails @ 6" o.c. 3" × 14 gage staples @ 6" o.c	Face nail
2. Ceiling joists to top plate	3-8d common (2-½" × 0.131"); or 3-10d box (3" × 0.128"); or 3-3" × 0.131" nails; or 3-3" 14 gage staples, ⁷⁄₁₆" crown	Each joist, toenail
3. Ceiling joist not attached to parallel rafter, laps over partitions (no thrust) (see Section 2308.7.3.1, Table 2308.7.3.1)	3-16d common (3-½" × 0.162"); or 4-10d box (3" × 0.128"); or 4-3" × 0.131" nails; or 4-3" 14 gage staples, ⁷⁄₁₆" crown	Face nail
4. Ceiling joist attached to parallel rafter (heel joint) (see Section 2308.7.3.1, Table 2308.7.3.1)	Per Table 2308.7.3.1	Face nail
5. Collar tie to rafter	3-10d common (3" × 0.148"); or 4-10d box (3" × 0.128"); or 4-3" × 0.131" nails; or 4-3" 14 gage staples, ⁷⁄₁₆" crown	Face nail
6. Rafter or roof truss to top plate (See Section 2308.7.5, Table 2308.7.5)	3-10 common (3" × 0.148"); or 3-16d box (3-½" × 0.135"); or 4-10d box (3" × 0.128"); or 4-3" × 0.131 nails; or 4-3" 14 gage staples, ⁷⁄₁₆" crown	Toenail[c]
7. Roof rafters to ridge valley or hip rafters; or roof rafter to 2-inch ridge beam	2-16d common (3-½" × 0.162"); or 3-10d box (3" × 0.128"); or 3-3" × 0.131" nails; or 3-3" 14 gage staples, ⁷⁄₁₆" crown; or	End nail
	3-10d common (3" × 0.148"); or 4-16d box (3-½" × 0.135"); or 4-10d box (3" × 0.128"); or 4-3" × 0.131" nails; or 4-3" 14 gage staples, ⁷⁄₁₆" crown	Toenail

(continued)

Source: the International Building Code, copyright © 2018, with permission from the International Code Council, Inc.

TABLE 2304.10.1–continued
FASTENING SCHEDULE

DESCRIPTION OF BUILDING ELEMENTS	NUMBER AND TYPE OF FASTENER	SPACING AND LOCATION
Wall		
8. Stud to stud (not at braced wall panels)	16d common (3-½" × 0.162");	24" o.c. face nail
	10d box (3" × 0.128"); or 3" × 0.131" nails; or 3-3" 14 gage staples, ⁷⁄₁₆" crown	16" o.c. face nail
9. Stud to stud and abutting studs at intersecting wall corners (at braced wall panels)	16d common (3-½ × 0.162"); or	16" o.c. face nail
	16d box (3-½" × 0.135"); or	12" o.c. face nail
	3" × 0.131" nails; or 3-3" 14 gage staples, ⁷⁄₁₆" crown	12" o.c. face nail
10. Built-up header (2" to 2" header)	16d common (3-½" × 0.162"); or	16" o.c. each edge, face nail
	16d box (3-½" × 0.135")	12" o.c. each edge, face nail
11. Continuous header to stud	4-8d common (2-½" × 0.131"); or 4-10d box (3" × 0.128")	Toenail
12. Top plate to top plate	16d common (3-½" × 0.162"); or	16" o.c. face nail
	10d box (3" × 0.128"); or 3" × 0.131" nails; or 3" 14 gage staples, ⁷⁄₁₆" crown	12" o.c. face nail
13. Top plate to top plate, at end joints	8-16d common (3-½" × 0.162"); or 12-10d box (3" × 0.128"); or 12-3" × 0.131" nails; or 12-3" 14 gage staples, ⁷⁄₁₆" crown	Each side of end joint, face nail (minimum 24" lap splice length each side of end joint)
14. Bottom plate to joist, rim joist, band joist or blocking (not at braced wall panels)	16d common (3-½" × 0.162"); or	16" o.c. face nail
	16d box (3-½" × 0.135"); or 3" × 0.131" nails; or 3" 14 gage staples, ⁷⁄₁₆" crown	12" o.c. face nail
15. Bottom plate to joist, rim joist, band joist or blocking at braced wall panels	2-16d common (3-½" × 0.162"); or 3-16d box (3-½ × 0.135"); or 4-3" × 0.131" nails; or 4-3" 14 gage staples, ⁷⁄₁₆" crown	16" o.c. face nail
16. Stud to top or bottom plate	4-8d common (2-½" × 0.131"); or 4-10d box (3" × 0.128"); or 4-3" × 0.131" nails; or 4-3" 14 gage staples, ⁷⁄₁₆" crown; or	Toenail
	2-16d common (3-½" × 0.162"); or 3-10d box (3" × 0.128"); or 3-3" × 0.131" nails; or 3-3" 14 gage staples, ⁷⁄₁₆" crown	End nail
17. Top plates, laps at corners and intersections	2-16d common (3-½" × 0.162"); or 3-10d box (3" × 0.128"); or 3-3" × 0.131" nails; or 3-3" 14 gage staples, ⁷⁄₁₆" crown	Face nail

(continued)

Source: the International Building Code, copyright © 2018, with permission from the International Code Council, Inc.

TABLE 2304.10.1–continued
FASTENING SCHEDULE

DESCRIPTION OF BUILDING ELEMENTS	NUMBER AND TYPE OF FASTENER	SPACING AND LOCATION
Wall		
18. 1" brace to each stud and plate	2-8d common (2-½" × 0.131"); or 2-10d box (3" × 0.128"); or 2-3" × 0.131" nails; or 2-3" 14 gage staples, 7/16" crown	Face nail
19. 1" × 6" sheathing to each bearing	2-8d common (2-½" × 0.131"); or 2-10d box (3" × 0.128")	Face nail
20. 1" × 8" and wider sheathing to each bearing	3-8d common (2-½" × 0.131"); or 3-10d box (3" × 0.128")	Face nail
Floor		
21. Joist to sill, top plate, or girder	3-8d common (2-½" × 0.131"); or floor 3-10d box (3" × 0.128"); or 3-3" × 0.131" nails; or 3-3" 14 gage staples, 7/16" crown	Toenail
22. Rim joist, band joist, or blocking to top plate, sill or other framing below	8d common (2-½" × 0.131"); or 10d box (3" × 0.128"); or 3" × 0.131" nails; or 3" 14 gage staples, 7/16" crown	6" o.c., toenail
23. 1" × 6" subfloor or less to each joist	2-8d common (2-½" × 0.131"); or 2-10d box (3" × 0.128")	Face nail
24. 2" subfloor to joist or girder	2-16d common (3-½" × 0.162")	Face nail
25. 2" planks (plank & beam—floor & roof)	2-16d common (3-½" × 0.162")	Each bearing, face nail
26. Built-up girders and beams, 2" lumber layers	20d common (4" × 0.192")	32" o.c., face nail at top and bottom staggered on opposite sides
	10d box (3" × 0.128"); or 3" × 0.131" nails; or 3" 14 gage staples, 7/16" crown	24" o.c. face nail at top and bottom staggered on opposite sides
	And: 2-20d common (4" × 0.192"); or 3-10d box (3" × 0.128"); or 3-3" × 0.131" nails; or 3-3" 14 gage staples, 7/16" crown	Ends and at each splice, face nail
27. Ledger strip supporting joists or rafters	3-16d common (3-½" × 0.162"); or 4-10d box (3" × 0.128"); or 4-3" × 0.131" nails; or 4-3" 14 gage staples, 7/16" crown	Each joist or rafter, face nail
28. Joist to band joist or rim joist	3-16d common (3-½" × 0.162"); or 4-10d box (3" × 0.128"); or 4-3" × 0.131" nails; or 4-3" 14 gage staples, 7/16" crown	End nail
29. Bridging or blocking to joist, rafter or truss	2-8d common (2-½" × 0.131"); or 2-10d box (3" × 0.128"); or 2-3" × 0.131" nails; or 2-3" 14 gage staples, 7/16" crown	Each end, toenail

(continued)

Source: the International Building Code, copyright © 2018, with permission from the International Code Council, Inc.

TABLE 2304.10.1–continued
FASTENING SCHEDULE

DESCRIPTION OF BUILDING ELEMENTS	NUMBER AND TYPE OF FASTENER	SPACING AND LOCATION	
Wood structural panels (WSP), subfloor, roof and interior wall sheathing to framing and particleboard wall sheathing to framing[a]			
		Edges (inches)	Intermediate supports (inches)
30. ⅜"-½"	6d common or deformed (2" × 0.113") (subfloor and wall)	6	12
	8d common or deformed (2-½" × 0.131") (roof) or RSRS-01 (2-⅜" × 0.113") nail (roof)[d]	6	12
	2-⅜" × 0.113" nail (subfloor and wall)	6	12
	1-¾" 16 gage staple, ⁷⁄₁₆" crown (subfloor and wall)	4	8
	2-⅜" × 0.113" nail (roof)	4	8
	1-¾" 16 gage staple, ⁷⁄₁₆" crown (roof)	3	6
31. ¹⁹⁄₃₂"-¾"	8d common (2-½" × 0.131"); or 6d deformed (2" × 0.113") (subfloor and wall)	6	12
	8d common or deformed (2-½" × 0.131") (roof) or RSRS-01 (2-⅜" × 0.113") nail (roof)[d]	6	12
	2-⅜" × 0.113" nail; or 2" 16 gage staple, ⁷⁄₁₆" crown	4	8
32. ⅞"-1-¼"	10d common (3" × 0.148"); or 8d deformed (2-½" × 0.131")	6	12
Other exterior wall sheathing			
33. ½" fiberboard sheathing[b]	1-½" galvanized roofing nail (⁷⁄₁₆" head diameter); or 1-¼" 16 gage staple with ⁷⁄₁₆" or 1" crown	3	6
34. ²⁵⁄₃₂" fiberboard sheathing[b]	1-¾" galvanized roofing nail (⁷⁄₁₆" diameter head); or 1-½" 16 gage staple with ⁷⁄₁₆" or 1" crown	3	6
Wood structural panels, combination subfloor underlayment to framing			
35. ¾" and less	8d common (2-½" × 0.131"); or 6d deformed (2" × 0.113")	6	12
36. ⅞"-1"	8d common (2-½" × 0.131"); or 8d deformed (2-½" × 0.131")	6	12
37. 1-⅛"-1-¼"	10d common (3" × 0.148"); or 8d deformed (2-½" × 0.131")	6	12
Panel siding to framing			
38. ½" or less	6d corrosion-resistant siding (1-⅞" × 0.106"); or 6d corrosion-resistant casing (2" × 0.099")	6	12
39. ⅝"	8d corrosion-resistant siding (2-⅜" × 0.128"); or 8d corrosion-resistant casing (2-½" × 0.113")	6	12

(continued)

Source: the International Building Code, copyright © 2018, with permission from the International Code Council, Inc.

TABLE 2304.10.1–continued
FASTENING SCHEDULE

DESCRIPTION OF BUILDING ELEMENTS	NUMBER AND TYPE OF FASTENER	SPACING AND LOCATION	
Wood structural panels (WSP), subfloor, roof and interior wall sheathing to framing and particleboard wall sheathing to framing[a]			
		Edges (inches)	Intermediate supports (inches)
Interior paneling			
40. ¼"	4d casing (1-½" × 0.080"); or 4d finish (1-½" × 0.072")	6	12
41. ⅜"	6d casing (2" × 0.099"); or 6d finish (Panel supports at 24 inches)	6	12

For SI: 1 inch = 25.4 mm.

a. Nails spaced at 6 inches at intermediate supports where spans are 48 inches or more. For nailing of wood structural panel and particleboard diaphragms and shear walls, refer to Section 2305. Nails for wall sheathing are permitted to be common, box or casing.

b. Spacing shall be 6 inches on center on the edges and 12 inches on center at intermediate supports for nonstructural applications. Panel supports at 16 inches (20 inches if strength axis in the long direction of the panel, unless otherwise marked).

c. Where a rafter is fastened to an adjacent parallel ceiling joist in accordance with this schedule and the ceiling joist is fastened to the top plate in accordance with this schedule, the number of toenails in the rafter shall be permitted to be reduced by one nail.

d. RSRS-01 is a Roof Sheathing Ring Shank nail meeting the specifications in ASTM F1667.

Source: the International Building Code, copyright © 2018, with permission from the International Code Council, Inc.

Termite Protection

Framers in many areas of the country have to be concerned about protection against termites. Pressure preservative-treated wood, naturally termite-resistant wood, or physical barriers can be used to prevent termite damage. The following map shows termite infestation probability by region.

Termite Infestation Probability

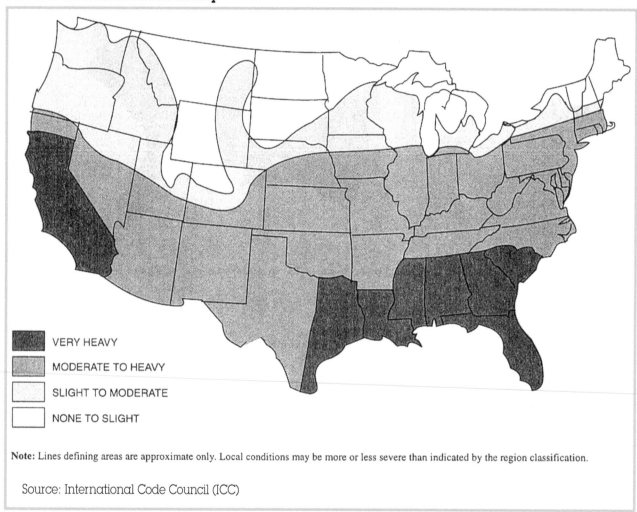

VERY HEAVY

MODERATE TO HEAVY

SLIGHT TO MODERATE

NONE TO SLIGHT

Note: Lines defining areas are approximate only. Local conditions may be more or less severe than indicated by the region classification.

Source: International Code Council (ICC)

Conclusion

An important part of your job as a lead framer is being aware of the building codes that apply to framing in your part of the country. You should be aware of how to use the code and of any revisions to those codes. Although locating information you need in the code books is often the hardest part of using the codes, the "Framing Index" at the beginning of this chapter should make this easier for you. It's a good feeling to know that you have framed a building the way it's specified according to code.

Chapter Eleven
GREEN FRAMING

Contents

Chapter Eleven

GREEN FRAMING

Green Framing is as much an attitude as it is an act of doing certain forms of framing. There are some framing designs and materials that are considered "green" but to be a green framer it takes a belief that you want to be a part of the global effort to reduce our effects on the ecosystem. In this chapter I will show you how the green movement is affecting the materials we use and in some cases the way we frame. I will also give you a basic understanding of the construction industry efforts to become green. In additions to this, I will discuss some behaviors that if you so choose will make you a part of the movement.

A little history is probably a good place to start. It's hard to say when the first discussions about how the human race is affecting the environment occurred and what we can do to our building practices to prevent ill effects, but Optimum Value Engineering, which has become known as *Advanced Framing*, was one of the first applications of green building. Advanced framing is an effort to conserve energy by altering framing techniques. It was soon realized, however that an overall building effort was needed to direct the construction industry in order to achieve the best effects. In 2000 the United States Green Building Council (USGBC) was formed and they created LEED (Leadership in Energy and Environmental Design), which is a construction and design industry joint effort to define and certify construction using green methods. The LEED program creates a tool for measuring the green building effectiveness by assigning credits in six areas: Sustainable Sites; Water Efficiency; Energy and Atmosphere; Materials and Resources; Indoor Environmental Quality; and Innovation and Design Process. Credits are totaled for individual jobs allowing for certification at different levels including, Certified, Silver, Gold, or Platinum. These certifications can be used in marketing programs.

LEED is an excellent construction- and design-oriented program to promote green building; however, it is a new and separate organization requiring its own fees and training. In 2008, the International Code Council (ICC) and the National Association of Home Builders (NAHB) came out with the National Green Building Standard ICC700-2008. Designed to guide the residential construction industry in green building, this standard was similar to the LEED system. It provided a rating system of environmental categories similar to LEED and performance levels of Bronze, Silver, Gold, or Emerald. The ICC700-2008 is a good guide, but is hard to regulate.

In 2010, the ICC published the International Green Construction Code (IGCC), the first ever compilation of international green building codes and standards. USGBC along with other agencies worked to help develop the IGCC. USCBC's LEED program set the format for guiding the design and construction industry in green building; however, it is a voluntary program and does not have jurisdictional enforcement capabilities.

The IGCC has similar topics for its five main content chapters; however, once the IGCC is accepted by a jurisdiction it becomes law for that jurisdiction. Unique to the IGCC for the other building codes is a section of regulations that relates to individual jurisdictions, so that each jurisdiction has to select from a group of regulations as to which ones they will require. There is an elective section where jurisdiction is required to determine the amount of a list of elective requirements that must be met.

It's all a bit confusing and most of it does not apply directly to framing; however, some items will. For example, Chapter 5 of the IGCC, "Material Resource Conservation and Efficiency," notes a requirement to develop a construction material and waste management plan that requires not less than 50 percent of non-hazardous construction waste to be diverted from landfills.

Green building is widespread in the construction process, but green framing is limited. Four parts of green framing that I will discuss are as follows:

1. Green Framing Feeling
2. Advanced Framing
3. Material Selection
4. Structural Insulated Panels (SIPs)

Green Framing Feeling

Green framing feeling sounds a little subjective, but that's because it has to be. For example, you are out on the job site framing, and the question is whether to throw a small cut off of a 2×6 into the trash or try to find a place to use it in your building process. Your decision is not only based on the cost of that piece of cut-off, but also the ease of just trashing it and the effect on the environment by using a new piece. Not an easy call to make, but you will have to make decisions like that all the time. If you choose the extra effort to conserve material, you will get a good green framing feeling.

Advanced Framing

A more tangible aspect of green framing is advanced framing. Based on the concept that wood is not as good an insulator as insulation, reduce the amount of wood in the exterior skin of a building and you will save energy and conserve building resources.

There are numerous ways to reduce the amount of wood in a building, but reducing the wood will reduce the building's strength. There are ways, however, to reduce the amount of wood that either don't affect the strength or still create strength enough to meet code requirements.

Some of the most common ways to reduce the amount of wood are the following:

1. Changing the stud layout from 16" O.C. to 24" O.C.
2. Changing common 3 stud backer to $2 \times \frac{1}{2} \times 6$ L backer or ladder blocking
3. Using drywall clips instead of wood backing
4. Using insulation in headers instead of wood fillers
5. Using a cripple header instead of solid headers for non-bearing walls
6. Using single top plates
7. Eliminating trimmers where not necessary
8. Eliminating window cripples
9. Adjusting layout or door and window locations so layout aligns with stud-trimmers
10. Changing the exterior wall from 2×4 studs to 2×6 walls.
11. Using standard lengths during building so that standard material can be used with less waste.

The "Advanced Framing" illustration shows these 11 techniques. They may already be integrated into your plans or you can integrate them on your own. If they are not already on your plans, make sure they do not conflict with the plans or that you receive the engineer's approval.

Advanced framing was originally developed to assist builders in using methods that would save energy in houses. Because energy conservation is a major component of green building, it is now a part of green framing.

Advanced Framing

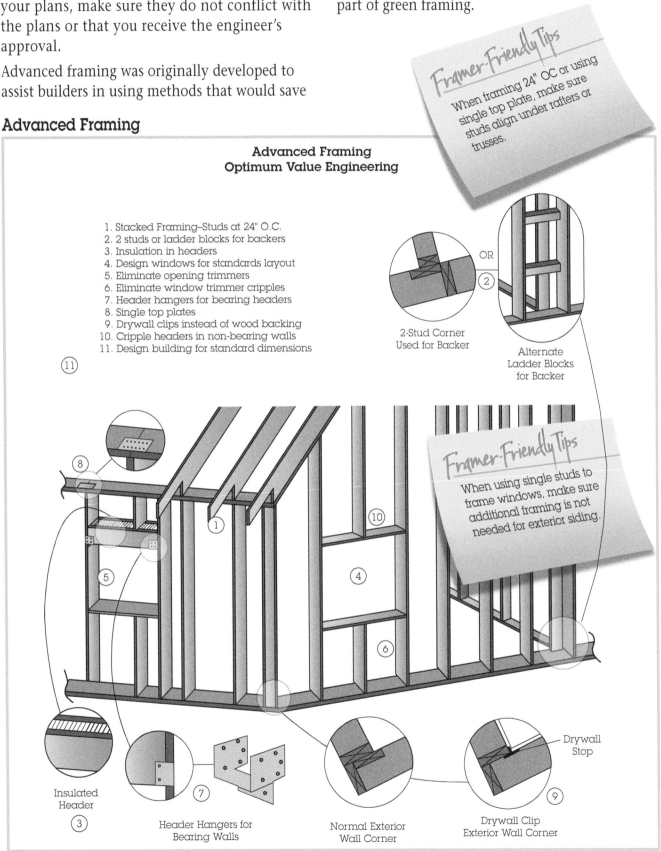

Advanced Framing
Optimum Value Engineering

1. Stacked Framing–Studs at 24" O.C.
2. 2 studs or ladder blocks for backers
3. Insulation in headers
4. Design windows for standards layout
5. Eliminate opening trimmers
6. Eliminate window trimmer cripples
7. Header hangers for bearing headers
8. Single top plates
9. Drywall clips instead of wood backing
10. Cripple headers in non-bearing walls
11. Design building for standard dimensions

Framer-Friendly Tips
When framing 24" OC or using single top plate, make sure studs align under rafters or trusses.

Framer-Friendly Tips
When using single studs to frame windows, make sure additional framing is not needed for exterior siding.

2-Stud Corner Used for Backer

OR

Alternate Ladder Blocks for Backer

Insulated Header

Header Hangers for Bearing Walls

Normal Exterior Wall Corner

Drywall Stop

Drywall Clip Exterior Wall Corner

Material Selection

Material selection is a concern for green framing. The following are six ways in which material selection is considered green.

1. **Use of Forest Stewardship Council (FSC) and Sustainable Forest Initiative (SFI) certified lumber.** This is lumber that is harvested following environmentally friendly guidelines for sustainable practices. This lumber is tracked through the chain of custody from the forest to the end user. It is labeled for identification.

2. **Use of salvaged or reused material.** Reusing lumber minimizes the need for new lumber.

3. **Use of regional material.** This is material that is harvested typically within 500 miles of the end use. The value gained is from the transportation energy savings.

4. **Use of rapidly renewable material.** This is usually considered material that has a 10-year or less growth cycle.

5. **Minimized use of volatile organic compounds (VOCs).** VOCs release toxins. Products such as subfloor adhesives will list the amount of VOCs that they contain. A tube of subfloor adhesive labeled "VOC Compliant" (see "VOC Compliant Subfloor Adhesive" photo) has VOC less water, less exempt solvent: <196g/l and <10.6% wt/wt. This makes it compliant with California ARB, which are among the strictest standards.

Forest Stewardship Council Label

Sustainable Forestry Initiative Label

Structural Insulated Panels (SIPs)

SIPs are a structural sandwich panel made of a foam plastic insulation core bonded between two structural facings usually made of oriented-strand board (OSB). SIPs are most commonly used for walls and roofs, but can also be used for floors and foundation systems. SIPs are considered green because of their expected energy savings. It is also expected there will be conservation of material because SIPs are made in shops.

If you are framing with SIPs, you will not be using the standard framing steps presented earlier in this book. Basically, you will be putting together panels of floors, walls and roofs with splines, headers, top and bottom plates, sealant, SIP tape, nails and SIP screws. Section R610, *Structural Insulated Panel Wall Construction,* of the International Residential Code gives prescriptive requirements for using SIPs for wall construction, but since each SIP manufacturer varies in their SIP construction it is best to follow the directions and plans the manufacturers provide.

SIP Installation

The concept of structural insulated panels was started in 1935, but it wasn't until 1990 that they were used enough for the Structural Insulated Panel Association to be created to organize the SIP industry. SIPs from different manufacturers are similar and their installation methods are common. SIP systems for floors, walls, and roofs can be used together, but are commonly used independently with standard framing for the other systems. The following installation steps give general directions for installing SIP walls. Walls are the most common use of SIPs and the installation of roofs and floors is similar. These steps can help organize the process.

1. **Organize your SIPs.** The most labor involved in erecting SIPs is moving of the panels. It is important that when the panels arrive on site that they are organized properly. If the SIPs are being set with a boom from the delivery truck then you need to coordinate with the SIP's manufacturer so that the panels are loaded in the sequence that you plan to use them. If the panels are unloaded from the delivery truck then you want to have your erection sequence developed and handy so you can unload and store the panels in a location that will make the first ones you need the easiest to access (see "SIPs Organized" photo).

SIPs Organized, Transit for Leveling, Sill Plate and Bottom Plate Attached

Sill Insulation

2. **Check the foundation or platform for level, square and dimensions.** When starting out it is important the walls have a level surface to sit on and the building is square. Where necessary, shim your bottom plates or use a power plane to make sure they are level, and when you are chalking lines for setting your bottom plates, adjust the lines so they are square and dimensioned per plan.

3. **Install sill plate.** If you are using sill insulation you will need to apply it before installing the sill plate. It is usually a piece of foam about ⅛" thick and the width of your sill plate. To install it just hold it in position over the anchor bolts and press down to punch a hole in the insulation (see "Sill Insulation" photo).

   The sill plate will probably need to be ripped to the dimension of the full width of the panel. For installation follow the same process used in standard wall framing. Mark and drill for your anchor bolts using your chalk lines for location and then align with the anchor bolts and drop into place.

4. **Install bottom plate.** The bottom plate needs to be the same width as the foam area of the panel, commonly 5-½". It will be bolted to the center of the sill plate so that the faces of the SIP will fit on either side. Both sides of the face will be nailed into the bottom plate. Place a bead of seal between the sill plate and the bottom plate. Secure the bottom plate and sill plate in place with a washer and nut on top of the bottom plate.

5. **Install splines where needed.** Sometimes the SIPs come with the splines attached on one side. If this is not the case you will need to insert the spline into the panel before you erect it. Check the panel to make sure that the area to receive the spline is free from debris, then apply continuous seal on the surfaces, insert the spline, and nail it in place (see "Attaching Spline" photo). Three common types of splines are shown in the *SIP Assembly* illustration later in this chapter.

6. **Check panel connection areas and electric chases.** Before standing the panel up, check

the one it will be connecting to. If the panel ends with an opening or a corner, check the location and trim if necessary. Circular saws and chain saws are commonly used for trimming panels and foam scoops can be used on the foam.

Where factory-supplied electrical chases are in the SIPs you will need to make sure that any splines you install have corresponding holes to allow for running electric wire. Mark and drill the splines if necessary before you install them.

7. **Seal panel to be installed**. Because SIPs are meant to be energy efficient it is particularly important to continuously seal all adjoining surfaces (see "Sealing" photo). The spline details in the *SIP Details* illustration later in this chapter show the locations of the seal.

8. **Set panel in place**. Tip the panel into place hinging on the far corner of the bottom of the wall (see "Installing SIP Panel" photo).

9. **Plumb and brace panel.** After the panel is standing check for proper placement, plumb both directions, and then nail in place and brace if needed. Sledge hammers, crow bars, long bar clamps, and come-alongs can be used to pull the panels together when needed. Allow a ⅛" space between panel faces.

Attaching Spline

10. **Nail and screw panel.** Panel screws come with the panels and you need to follow the manufacturer's suggested location for their installation. Also follow the manufacturer's instructions for size and

Sealing

Installing SIP Panel

spacing of the nailing. You will be using your standard framing nails. See the SIP Details illustration later in this chapter.

11. **Install top and double plates.** Top plates and double plates are installed to provide overlapping at intersections, corners, and splines. The top plate needs to be made of 2× wood, recessed into the panel, and nailed between the faces. The double plate needs to overlap the top plate a minimum of 2 feet and be the width of the SIP including the faces.

12. **Apply SIP tape.** As a last step, apply the SIP tape to the inside seams of all SIPs (see "Applying SIP Tape" photo).

The following *SIP Assembly* and *SIP Details* illustrations are from a SIP manufacturer and give you an idea of what you can expect for instructions from SIP manufacturers.

See "SIP Tools" photos for examples of tools that are commonly used for SIP installation, but not often used in standard framing. Standard framing tools are also used in installing SIPs.

Most important to SIP installations is organization. If you want to have a successful and productive job, do your homework and make sure you have all the tools you will need, your foundation is level and square, and you have a good understanding of where each and every piece of the puzzle will fit.

Applying SIP Tape

SIP Assembly

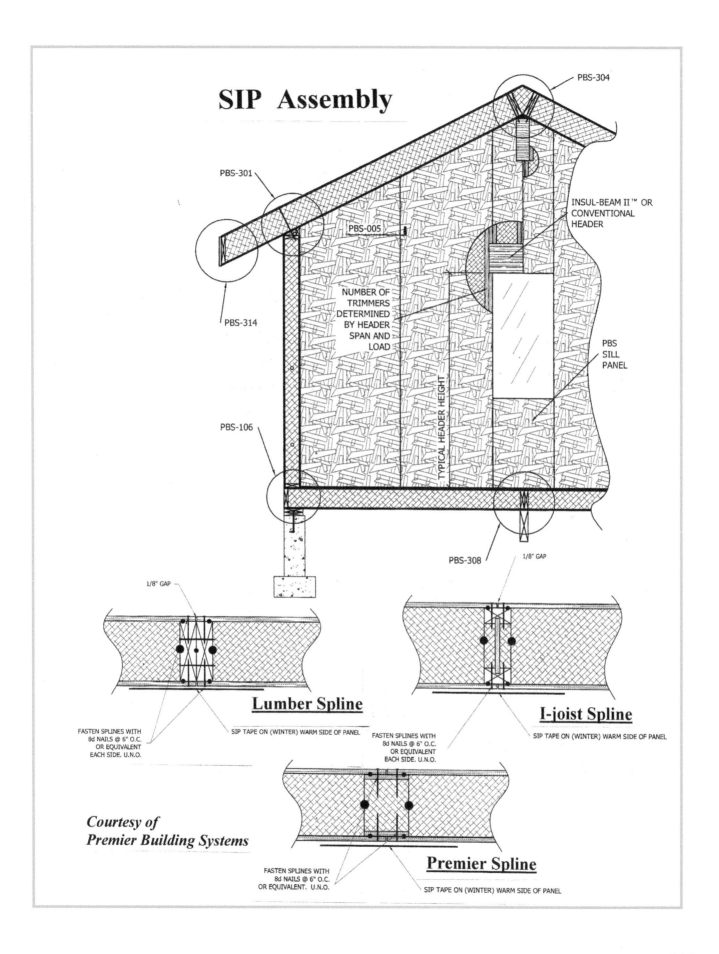

PBS-304

PBS-301

PBS-005

INSUL-BEAM II ™ OR CONVENTIONAL HEADER

PBS-314

NUMBER OF TRIMMERS DETERMINED BY HEADER SPAN AND LOAD

PBS SILL PANEL

TYPICAL HEADER HEIGHT

PBS-106

PBS-308

1/8" GAP

1/8" GAP

Lumber Spline

FASTEN SPLINES WITH 8d NAILS @ 6" O.C. OR EQUIVALENT EACH SIDE. U.N.O.

SIP TAPE ON (WINTER) WARM SIDE OF PANEL

I-joist Spline

SIP TAPE ON (WINTER) WARM SIDE OF PANEL

FASTEN SPLINES WITH 8d NAILS @ 6" O.C. OR EQUIVALENT EACH SIDE. U.N.O.

Premier Spline

FASTEN SPLINES WITH 8d NAILS @ 6" O.C. OR EQUIVALENT. U.N.O.

SIP TAPE ON (WINTER) WARM SIDE OF PANEL

Courtesy of
Premier Building Systems

SIP Details

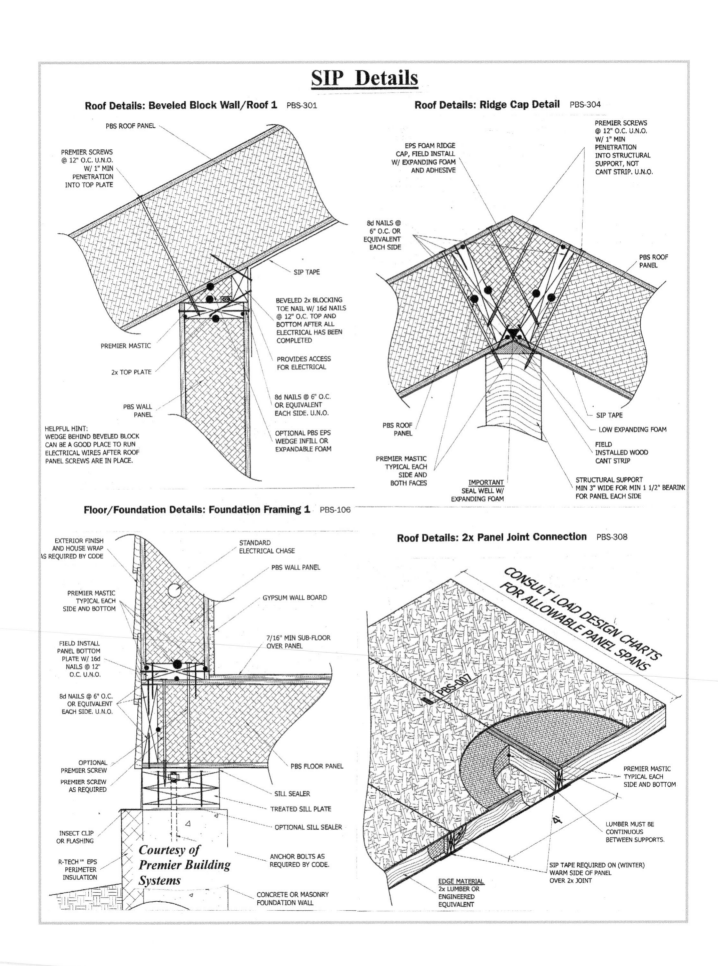

Roof Details: Beveled Block Wall/Roof 1   PBS-301

PBS ROOF PANEL

PREMIER SCREWS @ 12" O.C. U.N.O. W/ 1" MIN PENETRATION INTO TOP PLATE

SIP TAPE

BEVELED 2x BLOCKING TOE NAIL W/ 16d NAILS @ 12" O.C. TOP AND BOTTOM AFTER ALL ELECTRICAL HAS BEEN COMPLETED

PROVIDES ACCESS FOR ELECTRICAL

PREMIER MASTIC

2x TOP PLATE

8d NAILS @ 6" O.C. OR EQUIVALENT EACH SIDE. U.N.O.

OPTIONAL PBS EPS WEDGE INFILL OR EXPANDABLE FOAM

PBS WALL PANEL

HELPFUL HINT:
WEDGE BEHIND BEVELED BLOCK CAN BE A GOOD PLACE TO RUN ELECTRICAL WIRES AFTER ROOF PANEL SCREWS ARE IN PLACE.

Roof Details: Ridge Cap Detail   PBS-304

EPS FOAM RIDGE CAP, FIELD INSTALL W/ EXPANDING FOAM AND ADHESIVE

PREMIER SCREWS @ 12" O.C. U.N.O. W/ 1" MIN PENETRATION INTO STRUCTURAL SUPPORT, NOT CANT STRIP. U.N.O.

8d NAILS @ 6" O.C. OR EQUIVALENT EACH SIDE

PBS ROOF PANEL

PBS ROOF PANEL

SIP TAPE

LOW EXPANDING FOAM

FIELD INSTALLED WOOD CANT STRIP

PREMIER MASTIC TYPICAL EACH SIDE AND BOTH FACES

IMPORTANT
SEAL WELL W/ EXPANDING FOAM

STRUCTURAL SUPPORT MIN 3" WIDE FOR MIN 1 1/2" BEARING FOR PANEL EACH SIDE

Floor/Foundation Details: Foundation Framing 1   PBS-106

EXTERIOR FINISH AND HOUSE WRAP AS REQUIRED BY CODE

STANDARD ELECTRICAL CHASE

PBS WALL PANEL

GYPSUM WALL BOARD

PREMIER MASTIC TYPICAL EACH SIDE AND BOTTOM

7/16" MIN SUB-FLOOR OVER PANEL

FIELD INSTALL PANEL BOTTOM PLATE W/ 16d NAILS @ 12" O.C. U.N.O.

8d NAILS @ 6" O.C. OR EQUIVALENT EACH SIDE. U.N.O.

OPTIONAL PREMIER SCREW

PREMIER SCREW AS REQUIRED

PBS FLOOR PANEL

SILL SEALER

TREATED SILL PLATE

OPTIONAL SILL SEALER

INSECT CLIP OR FLASHING

R-TECH™ EPS PERIMETER INSULATION

ANCHOR BOLTS AS REQUIRED BY CODE.

CONCRETE OR MASONRY FOUNDATION WALL

Courtesy of Premier Building Systems

Roof Details: 2x Panel Joint Connection   PBS-308

CONSULT LOAD DESIGN CHARTS FOR ALLOWABLE PANEL SPANS

PBS-007

PREMIER MASTIC TYPICAL EACH SIDE AND BOTTOM

LUMBER MUST BE CONTINUOUS BETWEEN SUPPORTS.

SIP TAPE REQUIRED ON (WINTER) WARM SIDE OF PANEL OVER 2x JOINT

EDGE MATERIAL
2x LUMBER OR ENGINEERED EQUIVALENT

SIP Tools

12" Blade Beam Cutter

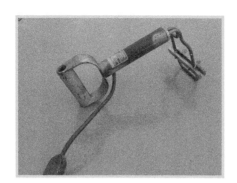

Foam Scoop

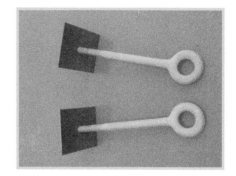

Lifting Eyebolts and Plates

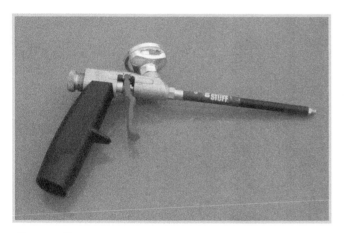

Foam Gun

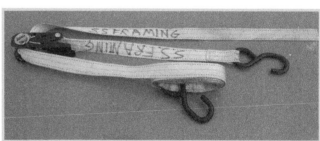

Ratchet Straps

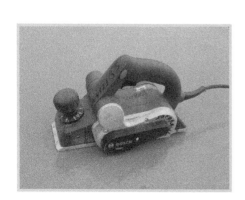

Power Planer

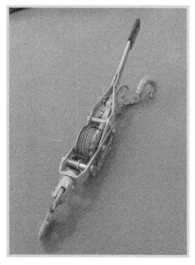

Come-along

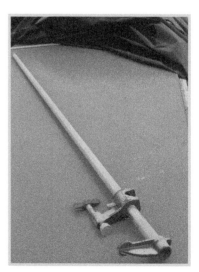

Bar Clamp

Chapter Twelve
PREPARING FOR A JOB

Contents

Chapter Twelve

PREPARING FOR A JOB

The best way to make any project start smoothly is to adequately prepare for the job. This means spending time looking over the plans, and organizing information, and talking with whoever is overseeing the job before you start working on the job site.

Often, the lead framer does this preparation the night or morning before a job starts. You'll find that the work will flow more smoothly if you begin preparation earlier and do it right.

If you're a carpenter working for a framing contractor or a general contractor, many of the preparation tasks listed in this chapter will be done for you. If, on the other hand, you are the lead framer, framing contractor, and home builder all in one, then it's up to you to get these done.

In this chapter, the word *superintendent* refers to the person on the job site who answers any questions related to the building. This person's actual title might also be *builder*, *owner*, or *framing contractor*. Although this book is about house framing, we use the word *building*, since the preparation is very similar whether it is a house, multi-family housing, a commercial building, or any structure where wood framing is used.

If you are preparing to start the job with the foundation slab in place, you will need to perform these four tasks:

1. Develop a job start checklist.
2. Review the plans and make preparations.
3. Organize the job site.
4. Conduct the pre-start job site review meeting.

11. **Hold-downs, Tie-downs, Anchoring System**

- It is best to install the hold-down studs when the wall is built, and it is easiest to drill the holes for the hold-down bolts before the hold-down studs are nailed into the wall.

- Have at least one hold-down of each size on the job site when you start. Because the hold-down sizes vary, it's good to have different sizes available so you can determine stud locations and bolt hole sizes and location. If you do not have the hold-downs, you can use a hardware catalog to determine hole sizes, locations, and stud locations.

12. **Truss Plans and Delivery Schedule**

- Many buildings have truss plans in addition to the plans provided by the architect. Because you want to line up the studs, floor joists, and roof trusses where possible, it is important to know where the truss manufacturer started the layout. You should use the truss layout and align the studs and floor joists. Truss plans typically call out where the layout starts.

- Often the truss plans are not drawn until shortly before they are needed. It is best to request the plans early so that they will be available when you need them.

- Check on the delivery date. Depending on the economy and the local truss manufacturers, the lead time for trusses can vary from days to weeks. You don't want to get to the roof and have to stop because the trusses aren't yet built.

13. **Steel Plans and Delivery Schedule**

- Typically if you have steel on the job, it should be in place before the wood framing is started. Check to see when it will be ready.

14. **Reference Point for Finish Floor**

- When you check the floor for level, it helps to have the benchmark used for the concrete work. If you don't have the benchmark, then you have to take a number of different readings to come up with an average before you can determine whether the concrete work is within tolerance. Sometimes the superintendent will be able to give you the benchmark.

15. **Reference Points for Wall Dimensions**

- Having the reference points will save you time in determining where the lines are actually supposed to be. Since the concrete work is seldom exactly where it is supposed to be, you will have to decide by how much the concrete is off and the best way to compensate for it without doing extra work or compromising the building.

- If you don't have reference points to work with, you will have to spend extra time taking measurements to determine where the mistakes are located in the concrete.

16. **Location of Job Site Truck**

- Be sure to locate your truck, trailer, or storage container close to the job site. Planning ahead with the superintendent can often open up a location that later could be occupied by other trades, material, or supplies.

Reviewing the Plans and Making Preparations

Plan review will save you time and energy, and make your work more productive. If you are framing a house with a plan you have used before, then you have already done the review. But if you are framing a new house design or, particularly, a multi-unit or commercial building, then it becomes very important to review the plans. Here are some of the most common ways of reviewing plans:

1. Study the plans. Sit down with the plans and figure out how the building is put together. Read the specifications. Most often they are standard and you can skim through them, but make sure to note anything that is new or different. Know enough about the new material so that you can understand the architect's explanations. If you can't figure it out, ask the framing contractor, superintendent, or architect about that particular element. If you are on a large job where the specifications come bound by themselves, you should know that they are probably organized under the Construction Specification Institute's (CSI) MasterFormat. Under this system, rough carpentry is listed in Division 6 as 06 10 00. This section contains the basic specification information about framing this job.

2. Make a list of questions. While you are studying the plans, have a pad of paper and pencil, or digital pad, handy so you can write down any questions. Go over these questions with the superintendent at the pre-start job site review meeting. Often, getting a question answered or a problem solved before the job begins saves an interruption in the framing. Even a little thing like the architect missing a dimension on the plans can cause a delay. If the superintendent okays scaling the missing dimensions, there won't be a problem; but if you need verification on missing dimensions, it's best to get them before you begin.

3. Highlight the plans. It's a big help to highlight easy-to-miss items on your plans. Use the same color highlights on all jobs so that it becomes easy to identify items for you and your crew. An example would be:
   Orange—Hold-downs
   Pink—Shear walls
   Green—Glu-lam beams
   Blue—Steel
   Yellow—Special items

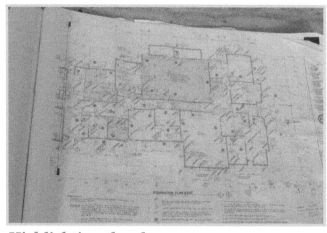

Highlighting the plans

4. Establish framing dimensions. Most rough openings are standardized, but because of exceptions and differences in floor covering, it's important to go over the rough openings before the job begins. The information sheets that follow can be used for reviewing these dimensions with the superintendent.

There is a sheet for 88-⅝" studs and one for 92-⅝" studs. These can be adjusted for different size studs. Go over each item with the superintendent or whoever is in charge of the job. Ask him/her to review the sheet and indicate that you will be using the rough-opening dimensions listed unless you are instructed differently. Note that 88-⅝" studs are standard because with a 4 × 8 header, they leave a standard 82-½" door opening. Note, too, that 92-⅝" studs work with a 4 × 12 header. (See "Standard Framing Dimensions" sheets.)

These sheets apply to residential framing. Commercial framing is not so standardized. Note that the use of hollow metal (H.M.) door and window frames is common in commercial framing. The frames are usually 2" in width. Rough openings (R.O.) for H.M. frames would typically be 2" for the frame plus ¼" installation space. As an example, a 3'-0" door would have an R.O. width of 3'-4-½", which is made up of 3'-0" for the door opening, 4" for the frames on each side, and ½" for the ¼" installation space on each side. The R.O. height would be 7'-2-¼", which would be made up of the 7'-0" for the door opening, 2" for the frame, and ¼" for installation space.

5. Tape the plans. Plan deterioration can be a problem, particularly at the end of a job. Use the same set of plans when possible so they include your highlighting and any changes that you have marked. When possible, request water-resistant print paper for the plans. If you're in a rainy area or season, this will keep the lines from running. Plastic covers are made to cover plans, but they can make it difficult to turn the pages. Clear plastic adhesive covering can be used, but then you can't write on the plans to note changes. A good system is to use clear plastic wrapping tape to tape the edges of the plans. This treatment usually provides the stability to make it through the job while still allowing for notes written on the plans.

Framer-Friendly Tips

Door and window manufacturers can provide specification sheets that show RO sizes.

Taping the plans

Standard Framing Dimensions 88⅝" Studs

		Header Size	Trimmer Size **
Stud height......................	88-⁵⁄₈"		
Wall height......................	93-¹⁄₈"		
R.O. windows...................	Width—nominal		
	Height—nominal	4 × 8	81-¹⁄₈"
		4 × 10	79-¹⁄₈"
R.O. exterior doors............	Width—nominal + 2-¹⁄₂"		
	Height—82-⁵⁄₈"	4 × 8	81-¹⁄₈"
	Height—82-¹⁄₈"	4 × 10	80-⁵⁄₈" cut T.P.***
R.O. sliding glass doors....	Width—nominal		
	Height—6'-10" door 82-¹⁄₈"	4 × 8	81-¹⁄₈" / ¹⁄₂" furr*
	82-¹⁄₈"	4 × 10	80-⁵⁄₈" cut T.P.
	Height—6'-8" door 80-¹⁄₈"	4 × 8	81-¹⁄₈" / 2-¹⁄₂" furr
	80-¹⁄₈"	4 × 10	79-¹⁄₈" / ¹⁄₂" furr
R.O. interior doors............	Width—nominal + 2"		
(nonbearing)	Height—82-⁵⁄₈"		81-¹⁄₈"
R.O. bifold doors..............	Width—nominal + 1-¹⁄₄" for ¹⁄₂" drywall		
	—nominal + 1-¹⁄₂" for ⁵⁄₈" drywall		
	Height—82-⁵⁄₈"		81-¹⁄₈"
R.O. bypass doors............	Width—nominal		
	Height—82-⁵⁄₈"		81-¹⁄₈"
R.O. pocket doors............	Width—2 × nominal + 1"		
	Height—84-¹⁄₂"		83"
Bathtubs........................	Width—nominal + ¹⁄₄"		
Tub fireblocks.................	14-¹⁄₂" from finish floor to bottom of block		
Medicine cabinet...........	R.O. 14-¹⁄₂" × 24"		
blocks	Height—48" from finish floor to bottom of R.O.		
	3" minimum away from wall corner		

These dimensions should be checked with the job site superintendent before beginning each job.

* Furr = furring under header after header is in place.

** Trimmer heights will increase by 1-¹⁄₂" if lightweight concrete is used or ¾" if gypcrete is used.

*** Cut T.P.—Cut the top plate out and leave the double plate.

R.O. (rough opening)—Any opening framed by the framing members.

Standard Framing Dimensions 92-5/8" Studs

		Header Size	Trimmer Size **
Stud height........................ 92-⁵⁄₈"			
Wall height..................... 97-¹⁄₈"			
R.O. windows................... Width—nominal			
Height—nominal		4 × 8	85-¹⁄₈"
		4 × 10	83-¹⁄₈"
R.O. exterior doors........... Width—nominal + 2-¹⁄₂"			
Height—82-⁵⁄₈"		4 × 8	85-¹⁄₈"/4" furr*
Height—82-⁵⁄₈"		4 × 10	83-¹⁄₈"/2" furr cut T.P.***
R.O. sliding glass doors.... Width—nominal			
Height—6'-10" door 82-¹⁄₈"		4 × 8	85-¹⁄₈"/ 4-¹⁄₂" furr
82-¹⁄₈"		4 × 10	83-¹⁄₈" /2-¹⁄₂" furr
Height—6'-8" door 80-¹⁄₈"		4 × 8	85-¹⁄₈"/ 6-¹⁄₂" furr
80-¹⁄₈"		4 × 10	83-¹⁄₈"/ 4-¹⁄₂" furr
R.O. interior doors............ Width—nominal + 2"			
(nonbearing) Height—82-⁵⁄₈"			81-¹⁄₈"
R.O. bifold doors.............. Width—nominal + 1-¹⁄₄" for ¹⁄₂" drywall			
—nominal + 1-¹⁄₂" for ⁵⁄₈" drywall			
Height—82-⁵⁄₈"			81-¹⁄₈"
R.O. bypass doors............ Width—nominal			
Height—82-⁵⁄₈"			81-¹⁄₈"
R.O. pocket doors............ Width—2 × nominal + 1"			
Height—84-¹⁄₂"			83"
Bathtubs......................... Width—nominal + ¹⁄₄"			
Tub fireblocks................. 14-¹⁄₂" from finish floor to bottom of block			
Medicine cabinet............ R.O. 14-¹⁄₂" × 24"			
blocks Height—48" from finish floor to bottom of R.O.			
3" minimum away from wall corner			

These dimensions should be checked with the job site superintendent before beginning each job.

* Furr = furring under header after header is in place.

** Trimmer heights will increase by 1-¹⁄₂" if lightweight concrete is used or ³⁄₄" if gypcrete is used.

R.O. (rough opening)—Any opening framed by the framing members.

Organizing the Job Site

After the plan review, you need to organize the job site. Figure out what your initial manpower needs and schedule are, and what tools you'll need for the job. The first day on the job site is usually a challenge.

1. **Manpower needs.** Typically, on the first day, your crew is ready to go to work and will be looking to you for instruction. At the same time you may not be sure if the concrete is level or the right size. Meanwhile, the superintendent may be on his way over with his list of things you need to take care of. If you have too many framers, everyone might be standing around until you get the job organized. If your schedule allows, start with just a two-man crew to check the foundation or slab for level and size and to get some lines chalked and some detailing done.

2. **Manpower tasks.** Knowing which jobs you want each framer to do before you get there always helps. Also, keep a couple of back-up tasks (such as cleaning out the truck or fixing tools) in mind in case something prevents you from starting right away. First-day jobs might include:

   - Cleaning the slab or foundation
   - Checking concrete dimensions
   - Checking level of concrete
   - Cutting makeup and headers
   - Nailing makeup and headers
   - Chalking lines
   - Setting up chop saw (radial arm or similar)
   - Building plan shack
   - Detailing plates

3. **Tools.** Not having the right tools can be like trying to cut the Thanksgiving turkey with a table knife. The tool list that follows will help you determine what you need. For example, you can look at the plans to find out what size bolts are being used so you can be sure to have the appropriate drill bits and impact sockets ready.

   It's easy to show up the first day without some of the necessary tools. Also, you might use different tools at the beginning of a job and at the end of a job. Highlighting the tools you need on the Tool List before the job starts will help you prepare and save time.

   Note that the "Location" column on the Tool List at the end of this section refers to the location where the tools are kept. (See legend on tool list.) The locations listed can be adjusted to your own situation.

4. **Plans.** Any time you can devote to the plans before you start the job is probably well spent. Two things are particularly important for getting started. First, decide where you are going to pull your layout from (see Chapter 7), and second, decide which lines you are going to set for reference (see "Getting Started" in Chapter 13).

   Looking at plans on the job site can be like trying to read a map while on a motorcycle: there is always the sunshine, wind, or rain. On the job site, you'll be juggling a number of things. Your crew will be asking you what to do next, and you'll have to think about the material you need and if you have enough nails, for example. It will take you about an hour to absorb as much information from the plans on the job site as you can in fifteen minutes off the job site. A good habit is to review the plans for ten minutes every morning away from the job site. You'd be surprised at how many mistakes are avoided by doing this.

Tool List

Tool	Location	Quantity
Framing saw	SB	1 per framer
Saw blades	H	Many
Cut saw	TB	2 per crew
Cut saw blades	FB	5 per crew
Impact wrench	TB	1 per crew
Impact sockets		
$3/8$" for SDS $1/4$"	MB	1 per crew
$3/4$" for $1/2$" bolt	MB	1 per crew
$15/16$" for $5/8$" bolt	MB	1 per crew
1-$1/8$" for $3/4$" bolt	MB	1 per crew
1-$5/16$" for $7/8$" bolt	MB	1 per crew
1-$1/2$" for 1" bolt	MB	1 per crew
1-$13/16$" or 1-$7/8$" for 1-$1/4$" bolt	MB	1 per crew
Drill	TB	2 per crew
Drill bits		
$5/8$"	FB	2 per crew
$3/4$"	FB	2 per crew
$7/8$"	FB	2 per crew
Router	TB	1 per crew
Router bits		
Panel pilot	FB	2 per crew
$1/2$" round	FB	1 per crew
Router wrench set	MB	1 per crew

Tool	Location	Quantity
Chop saw	TB	1 per crew
Beam saw	TB	1 per crew
4-way electric cord	H	1 per crew
100' electric cord	H	1-$1/2$ per framer
Nail gun	LB	1 per framer
GWB nail gun	LB	1 per crew
Air compressor	TB	1 per three framers
100' air hose	H	1-$1/2$ per framer
2' level	TT	1 per crew
4' level	TT	1 per crew
8' level	TT	1 per crew
Sledgehammer	FT	2 per crew
Crowbar	FT	2 per crew
Framing square	TT	1 per crew
Stair nuts set	FB	1 per crew
Glue gun	TB	2 per crew
Wall pullers	TB	2 per crew
Hand saw	H	1 per crew
Transit stand	FT	1 per crew
Transit	LB	1 per crew
100' tape	FB	1 per crew
String line	TB	2 per crew
Water jug	T	1 per crew
Step ladder	T	1 per crew
Extension ladder	T	1 per crew
First aid kit	FT	1 per crew
Microwave	FT	1 per crew
Stereo	T	1 per crew

Legend:

SB = Saw Box

H = Box

FT = Front of Truck

TB = Tool Box

SR = Screwdriver Rack

JH = Jay Hooks

MB = Metal Box

T = Truck

LB = Lock Box

FB = Flat Box

TT = Top of Truck

Tool List (continued)

Tool	Location	Quantity
Broom	T	1 per crew
Chalk bottle	T	1 per crew
Knife blades case	H	1 per crew
Vice grip	MB	1 per crew
5" crescent wrench	MB	1 per crew
8" crescent wrench	MB	1 per crew
Allen wrench set	MB	1 per crew
Screwdriver		
Standard	SR	2 per crew
Phillips	SR	2 per crew
PLUS		
Retractable safety line	JH	2 per crew
Lanyards	JH	4 per crew
Regulators	FB	½ per gun
Compressor oil	TB	1 per crew
Gun oil	TB	1 per crew
Plumb bob	FB	1 per crew
Electric three-way	FB	2 per crew
Air three-way	FB	2 per crew
Saw guides	FB	1 per crew
Screwdrivers	FB	Misc.
Chain saw	SB	1 per crew
Chain saw blades	FB	1 per crew
Palm nailer	LB	1 per crew
Ear plugs	FB	Misc.
Back support	JH	Misc.

5. **Schedule.** Developing a schedule is a difficult task, and one that should be a responsibility of the contractor. If, however, the framing contractor does not provide one, the lead framer should create his own. It is a valuable tool that will help you organize the job and then analyze how the work is going.

6. **Plan shack.** On bigger jobs, a plan shack is a good tool to have. It doesn't have to be fancy, but if it keeps your plans dry and helps keep the job organized, it is worth the time and material.

Plan shack

Conclusion

Your time spent preparing for a job sets the tone for managing the whole job. It lets you hit the job running and puts everyone on notice that you are serious about making this job run smoothly.

With a picture of the plans in your head, a job site check list complete and your tools organized, you will start out answering questions and taking control of what needs to be done to get your project framed.

Chapter Thirteen
MANAGING THE FRAMING START

Contents

Chapter Thirteen

MANAGING THE FRAMING START

Certain activities must take place before you begin framing. The dimensions and level of the foundation and slab need to be checked. If they are not perfectly level (which is not unusual), you must determine how far from correct they are, whether they are within tolerances, and what types of adjustments you must make. It is important that the dimensions are accurate, and the building is square before you start. Note, too, that the cabinets, floor covering, drywall, roof trusses or rafters, and much more depend on the measurements being accurate and square.

The four steps to getting started, covered in this chapter, are:

1. Checking the exterior wall dimensions
2. Checking the reference lines for square
3. Adjusting the reference lines to correct dimensions and square
4. Checking the building for level

Checking Exterior Wall Dimensions

If you have the concrete-work reference points handy, getting started will be easier for you. If you don't, establish reference lines of your own. Be sure to mark these lines well, since you will be using them throughout the job. Using clear marking paint in inverted cans makes it easy to protect your lines on the concrete.

You will want to use the reference lines to find any deviations from plan measurements or any out-of-square parts of the foundation. Start by stringing dry lines that will allow you to measure. The more of the building you can measure from these lines,

the more likely you are to find any mistakes. Look at the plans, and string two dry lines perpendicular to each other and covering as long a distance of the building as possible. If you can add two more dry lines, one on each side and opposite to the first two, that will help. (See the "Start-Up" example.) Once you have established your lines, take measurements between the lines and to the major exterior walls in the building. Make a quick footprint of your building, and as you measure the distances, write them down on the footprint. (See "Footprint Sketch Dimensions" example later in this chapter.)

A laser can also be used to establish square lines. The laser will give you dots that you just need to connect. Set the laser up at a convenient position so you will be able to chalk reference lines. Once you have established the lines, use your tape and a 3-4-5 triangle to check for accuracy.

"Laser Dots." Note: a fifth laser dot would be visible under the back of the laser.

Checking Reference Lines for Square

To check using four reference lines, measure the two diagonals, then write down the measurements on the footprint you used for measuring the dimensions. If the corner points are set correctly, then the diagonals will be the same length. If the reference dry lines are square, then the diagonals will be the same length. (See "Start-up" example.)

If you have only two reference lines to work with, you'll need to use a triangle to help you check for square. The two reference lines will be "square" with each other if they create a right angle (90°). You can use a 3-4-5 triangle or the Pythagorean

To check the exterior wall dimensions:

1. String dry lines to create reference lines.
   - Select lines that are as long as possible.
   - Locate line ends at extreme ends of the building.
   - Locate lines so that they reference the entire building.
2. Use reference lines to measure the building dimensions.
   - Check the exterior wall locations and note any discrepancies.

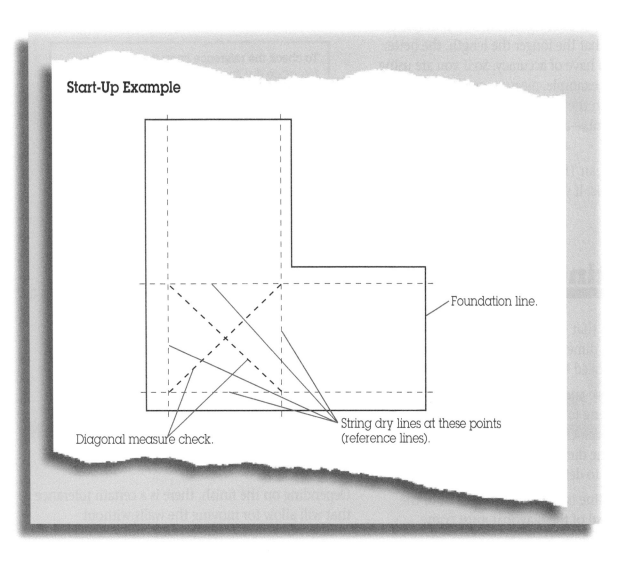

Start-Up Example

Foundation line.

String dry lines at these points (reference lines).

Diagonal measure check.

The "Footprint Sketch Dimensions" illustration is made on the job site. It will help determine how to best adjust your reference lines to make the building square. In this example, four dry lines are established to form a square. The diagonal distances that should be the same are then checked. Because they are different, the reference lines will need to be moved to make the diagonals the same. By comparing the actual and the planned dimensions of the walls that the reference lines are measured from, you can determine which reference lines should be moved. When you move a reference line, the other lines are affected.

If you have all the information down on your footprint sketch, you can come pretty close to knowing exactly how much to move each line, and keep making adjustments until you are comfortable with your accuracy. Once your reference lines are established, you can set all the other lines in the building from them. The measurements in circles on the sketch show the distance that the reference lines would be first moved. It is difficult to determine exact amounts because of the proportions, but if you study the footprint for a little while, you can come pretty close.

Checking level using a rotary laser.

Checking the Building for Level

Using a transit or laser is the best way to check for level. A water level can also be used. Once a level foundation has been established, you are ready to cut, drill, and set the mudsill in place.

The foundation and/or slab should be ready for you to start framing when you first arrive. Sometimes, however, this is not the case, and time will be needed to "shoot" (measure using a transit or laser) a foundation and slab. Time must also be allotted to fix any problems in the concrete. It will be your responsibility to check and make a suggestion if you think corrective work is necessary. Start by checking and recording your findings. Record your findings in a way that lets you use the information if you decide the concrete needs corrective work. To record your findings, make a footprint sketch similar to the one you used for dimensions and squaring, and write the readings on the footprint. (See "Footprint Sketch Elevations" later in this chapter.)

To take the measurements using a transit, one framer should hold a tape measure at the spots to be measured, while another framer uses the transit to record the height to the transit line from the concrete. To take a measurement using a rotary laser, one framer records the measure at the spots to be measured using a detector that reads the laser beam.

If you are working with a foundation wall or an existing wall, a rotary laser is efficient because once you have the laser set up, you can just mark the red line and measure up or down from it. If the concrete work is done well, typically within a variance of ¼", then just shooting at strategic locations on the concrete should be sufficient to check for level. If you quickly find out that the concrete is not level, you will need to shoot the concrete every four to eight feet along the walls. Either way, be sure to record the measurements on the footprint sketch. Mark the locations where the measurements were taken. When you start building walls, you will use the measurements in the footprint and the marks on the concrete to determine stud heights. The marks are only made every 4'–8', because when you are laying out walls, a level can be used to find the heights between the marks. Another way to find the stud heights between marks is to use a chalk line at the top of the wall to rub studs against and mark the heights. (See "Chalk Line at Top of Wall" photo.)

Once you have finished a footprint with the elevations marked, you can determine if any corrections need to be made. With the elevations written down, you can show the footprint to the superintendent or owner to let them decide what tolerance they will accept on their building.

If you look at the "Footprint Sketch Elevations" illustration, you will notice that most of the building elevations center around 49-¼" and are within ¼". The top wall on the sketch, however, appears to be low, with the lowest point at 49-⅝". Although 49-⅝" is more than 49-¼", it actually represents a low point, because the measurement represents the distance from the transit line down to the concrete.

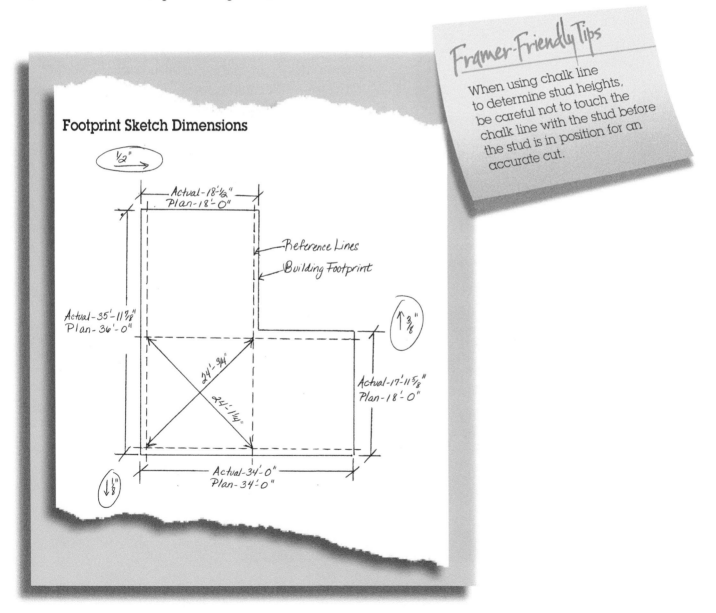

Footprint Sketch Dimensions

Framer-Friendly Tips

When using chalk line to determine stud heights, be careful not to touch the chalk line with the stud before the stud is in position for an accurate cut.

On the footprint sketch example, you would probably want to use a height of 49-¼" and recommend adjusting the section of the building that is low.

The Xs on the footprint represent the position of your tape measure when you shoot the height with the transit. Mark the X on the concrete so that when you start building walls, you will have a reference point if your heights need adjustment. Also keep your footprint sketch for this purpose.

Finding Stud Heights for Different Height Foundations

Finding stud heights when the slab or foundation is not level is difficult. It is even more so when the foundation steps up or down to different heights. Using a transit requires you to measure everything from the height that is established when you set the transit. It could be measuring up or down from the transit reading line. It is hard to keep all the numbers in your head when you start adding and subtracting for the different concrete levels and the levels of the foundation.

The best way to find the individual stud heights is to write everything you need for each individual stud down on a piece of paper and figure the stud height from those figures. The illustration "Stud Heights for Different Foundations" illustrates how to do this and includes a "Job Site Worksheet." Whenever you move the transit, it changes the measurements, so you would have to start over if you did not have all the measurements you need for a particular area. It is easier to finish one area completely before moving the transit.

Chalk line at top of wall

Transit line from concrete

When there is a foundation wall that is above a slab, it saves time if you chalk a level line on the foundation and then measure the stud heights from this line.

Conclusion

Before you begin framing, it's crucial to check your foundation and slab to make sure they are square. Once you have done this, you can move to the next step, the actual framing.

Framer-Friendly Tips

If the anchor bolts are bent, a piece of cast iron pipe works well to straighten them without banging up the threads or damaging the concrete. Slip the pipe over the bolt, and push until straight. The longer the pipe, the easier the push.

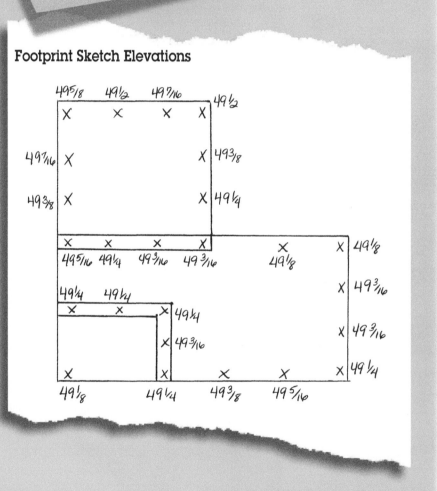

Footprint Sketch Elevations

Stud Heights for Different Foundations

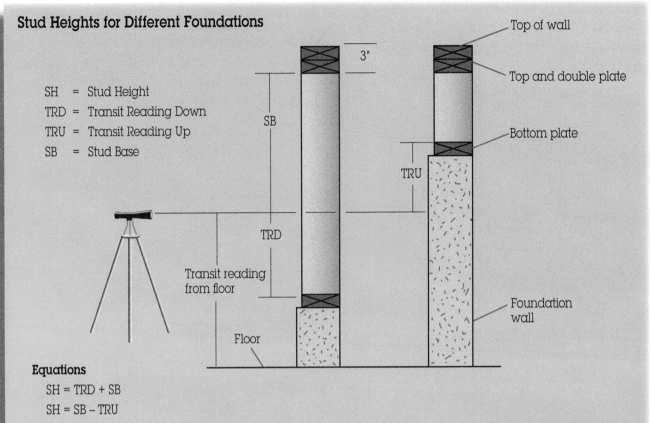

SH = Stud Height
TRD = Transit Reading Down
TRU = Transit Reading Up
SB = Stud Base

Equations

SH = TRD + SB
SH = SB – TRU

Steps

1. Set the bottom plate.
2. Find SB (stud base) by looking on the plans for height of wall; then subtract the height of the "Transit reading from floor" and the 3" for top and double plate.

Job Site Worksheet			
TRD	+	SB =	SH
SH1	+		=
SH2	+		=
SH3	+		=
SH4	+		=
SH5	+		=
SH6	+		=

	SB	–	TRU	=
SH7		–		=
SH8		–		=
SH9		–		=
SH10		–		=
SH11		–		=
SH12		–		=

Chapter Fourteen
MANAGING A FRAMING TEAM

Contents

Chapter Fourteen

MANAGING A FRAMING TEAM

This chapter is intended for advanced framers who are becoming "lead framers," or starting to manage a framing crew. Keep in mind that the lead framer's productivity is defined by the productivity of the crew. If you're taking on the job of lead framer, you'll need to think about the information your crew needs and how to teach and manage them most effectively. Earlier chapters in this book will help you train your crew in the specific steps of various framing tasks. But there are other aspects to managing a crew. It's the lead framer's job to get the building framed on time and within budget. The lead framer must also be sure that the expected quality standards are met, and that the building is structurally sound, visually aligned, and ready for inspectors and for other trades—all of this while maintaining a safe and congenial workplace. To meet all of these goals is an impressive accomplishment. The purpose of this chapter is to help you get there.

Management techniques have been developed over the years by studying and applying methods that work. The trend has been away from the dominating "command" approach and toward the cooperative "team" approach. This chapter deals with some organizational tasks, as well as with relationships and motivation. Developing good working relationships and instilling motivation is probably the most important and the most difficult task of a leader. A construction project manned by crews of skilled craftsmen who take pride in their work and get along with each other is bound to be successful. Assembling and directing such crews can only be accomplished by a leader who has developed good management skills.

Managing a framing team is a task like no other. The job changes every day and is always full of new surprises. The lead framer should be good at multi-tasking. A typical day might include trying to make sense of plans that don't provide enough information, dealing with an owner or general contractor who is focused more on cost and schedule than the details of framing, and organizing a group of framers who have different levels of knowledge and experience into an effective team.

The most valuable tools you can have in managing a crew are common sense, framing knowledge, the ability to evaluate a situation objectively, and an understanding of your crew's abilities and personalities. You probably already have a preferred management style, based on what you have learned in your experience in the field. This chapter will help you better understand that management style and improve upon it.

Managing Your Team

Construction is a unique industry. It is always changing. Each new job or building has its own individual plan, timetable, and workers to do the job. The economy, local governments, codes, tools, and materials are also constantly changing, creating different work environments. The crew structure has to change, as necessary, to accommodate the particular requirements of a job. To be efficient, the lead framer must be aware of all factors that affect the job, and must be able to work successfully within them.

The management structure of a framing crew can differ from company to company. In some cases, the lead framer is the owner/builder. In larger construction companies, the lead framer may run only the framing crew. Either way, the lead framer leads the framing on the job.

This book is not intended to cover the functions of the framing contractor or builder that include office management, bids, payroll, or business organization. It is written for someone who already has experience, knowledge, and skills in basic framing, and who wants to move up to the next level or become a better lead framer.

As a lead framer, you need to have a different perspective from a crew member. When you are working on your own, the amount of work completed depends on you. When you are leading a crew, the amount of work finished depends on the whole crew. On your own, you have complete control over what can be done, whereas you have limited control over how much work your crew gets done. Nevertheless, you only need a little control and increased knowledge to make a big difference in how much work the crew finishes. This chapter is intended to give you that increase in knowledge—which can make your jobs run better.

The lead framer

The Role of Lead Framers

A Lead Framer Must Possess:

1. Knowledge to frame any building
2. Ability to impart knowledge to other framers
3. Ability to motivate other framers

Knowledge to Frame Any Building

As a lead framer, you must thoroughly understand the basic concepts involved in framing any style building. The framing crew takes their direction from you; you, in turn, take your direction, depending on the situation, from any of the following:

a. Framing contractor
b. Site superintendent
c. Architect or engineer (plans)
d. Owner

There are a number of framing requirements that are easy to overlook. Compile checklists such as the ones shown and refer to them during each phase of the job.

Framing Checklist

Walls:
- ☐ Studs for tubs, medicine cabinets, etc.
- ☐ Special shear nailing
- ☐ Hold-downs
- ☐ Alignment of windows and doors
- ☐ Size of studs

Joists:
- ☐ Heading-off for toilets
- ☐ Double under bearing walls
- ☐ Cantilevering overhangs

Roof:
- ☐ Fascia overhang 6" for gutters
- ☐ Skylights or roof hatch
- ☐ Ventilation
- ☐ Attic access

Crawl space:
- ☐ Vents in rim joists
- ☐ Crawl space access

Responsibilities of Lead Framer

- ☐ Check location and quality of power supply.
- ☐ Check location and date of lumber drop.
- ☐ Check window delivery schedule.
- ☐ Check truss delivery schedule when appropriate.
- ☐ Arrange to have the builder complete as much site preparation as possible before starting, including leveling the area around the building where framers will be working.
- ☐ Highlight items on plans that are easy to miss or hard to find.
- ☐ Make a list of potential problem areas and items that are easy to forget.

Ability to Impart Knowledge to Other Framers

- When teaching someone, start with the basics. Assume nothing. Explain in clear and simple language exactly what the job is, and how it is to be done.

- The easiest way to lead may be to give orders, make demands, and threaten. However, it creates an unsettling atmosphere that is not conducive to a cooperative, self-motivated crew. *Request* that framers do tasks; do not *order* them.

- Assume that no framer intentionally does something wrong. Help your crew correct errors and show them how to avoid making them again.

- Treat each framer with respect. His time may be less valuable to the company, but his worth as an individual is equal to yours.

- The words "please" and "thank you" can make a framer feel much better about working for you. It is an easy way to let him know that what he does is important and appreciated.

- Do not give the hard, unpleasant jobs to the same framer time after time. The entire crew should share such tasks.

- When a framer asks you a question, give him the answer, but then explain how you got the answer so the next time he can figure it out himself.

Ability to Motivate Other Framers

To produce good work efficiently, a framer must be motivated. To be motivated, a framer must:

- Feel good about himself
- Feel what he is doing is important
- Be respected by his lead framer
- Feel he is being treated equally

Feel good about himself

You are a lead framer, not a therapist, but your attitude toward your crew should have a positive effect on his motivation. A crew whose members take pride in their individual and collective skills will invariably produce quality work and take pleasure in doing it.

Feel what he is doing is important

Every task, no matter how small, is necessary to complete the job and, therefore, important.

Be respected by his lead framer

Take time to listen and teach. If, as lead framer, you are called upon to solve a framing problem, it is better to let the framer explain his solution first and, if it is an acceptable solution, let him do it his way. There are often several ways to solve a framing problem. If you have a way that is much faster or easier than the framer's way, explain it to him and tell him how you came to your conclusion.

Directions should be given in terms of the job, not the individual. For example: *Negative*—"I told you five minutes ago to build that wall." *Positive*—"We need that wall built right away so we can finish this unit."

Framers like to feel that the person supervising is concerned about what they think and how they feel. Convey this through your words and actions.

Feel he is being treated equally

Don't show favoritism when assigning tasks. Make every effort to treat all framers fairly. Deal with any complaints impartially.

Different Types of Management

There are many ways to manage framers. It is important to know the different management styles and the effects that they have on employees, so that you can create the most productive framing crew. There are three main styles of management: *autocratic, bureaucratic,* and *democratic.*

Autocratic: The lead framer has the decision-making power and does not delegate authority. Discussion and suggestions are generally not permitted. This style sometimes motivates framers to please the lead framer instead of improving productivity. It also discourages framers from finding creative solutions.

Bureaucratic: The lead framer enforces established rules, regulations, policies, and procedures to run the crew. This style does not allow for creative solutions.

Democratic: Framers help determine the goals of the company. The lead framer organizes and directs the framers as part of his job as a crew member. If a problem comes up between the lead framer and a framer, they work together to find a solution they can both accept. This style usually creates a congenial work atmosphere.

Most lead framers use a combination of the three styles. This lets them have authority when they need it, while getting the help from their framers in developing the most productive methods for accomplishing their work.

Managing a framing crew can be compared to playing quarterback on a football team. Your team has to have confidence in your ability to direct them. The team expects you to tell them when they make mistakes, but they also expect you to tell them what they need to know to do a good job and to be considerate of them. In effect, you have to develop a working relationship with each framer.

The autocratic lead framer

The democratic lead framer

Different Types of Framing Crews

There are different types of framing crews, which require some adjustment in style. They are:

1. Hourly employees
2. Piece workers
3. A combination of the two

Hourly workers are paid by the hour. Their goal is to keep their employer happy with their work. They typically are more concerned with quality of work than with speed. Piece workers are paid by the amount of work they finish. Their main goal is to get as much work as possible done within a certain time frame. A combination of the two allows for the employee to be paid for each hour he or she works, then to receive a bonus for completing extra work within a defined time frame. A combination system can provide the motivation to maintain speed, while still allowing you a great degree of control over the job.

More on Motivation

Motivation is the intangible factor that can make or break a crew, and probably the single most important factor that affects framers, yet it is not something you can demand of your crew. As a lead framer, you want to support individual framers and maintain a high level of motivation in the crew.

Ideas for Building Relationships and Motivation

1. Honesty is a basic. A framer will observe not only what you say and how you treat him or her, but what you say to others and how you treat them. Keep your framers well-informed. If there is a slow-down coming up, and some framers might be laid off, let them know. You risk the chance of them quitting before the job is finished, but if you want them to be on your side, you have to be on theirs.

2. The first day on a job is the most important time for setting a new framer's attitude toward his or her job. Take time to introduce him to the whole crew and show him where he can find tools, the first aid kit, and portable toilets. Allow time for him to acclimate to the job.

3. Developing relationships takes time and a conscious effort. While your time is valuable, and you have to balance it, try to listen to what your framers have to say, and show patience. If you want them to support your interests, you have to be concerned about theirs. Make yourself available and easy to talk to. Encourage open and free resolution of problems, and make every effort to use your framers' suggestions, or explain why if you decide not to. This gives the framers positive feedback and gets them thinking about better and faster ways to accomplish tasks. If you constantly reject their suggestions, you reduce their motivation.

4. Use power discreetly. The more you have to display authority, the less valuable it becomes. Persuasion and guidance can be more effective than a show of authority.

5. Assign more responsibility and train framers to take on new tasks whenever the job allows. This will motivate framers to take on more duties.

6. Teach framers how to solve problems.

7. Praise framers for good work. This helps create a positive attitude, especially when it is done publicly. Compliments are a good relationship-builder, especially when framers first start working with you. Go out of your way to find something they have done well. Hopefully you can get a couple of positive compliments in before you have to start pointing out any mistakes.

8. Expect some mistakes and use them as learning opportunities. Making mistakes and learning on the job are everyday occurrences and should not create fear in a framer. Your framers need to know that you are there to teach and direct them, and that you will be fair and reasonable.

9. Make criticism into a learning experience and give it in private. When a framer makes mistakes or is sloppy, don't assume it's intentional. Calmly explain what he did wrong. Direct the criticism at the action, not the person. Be specific. For example, "Your nails are not sunk deep enough," instead of, "You are a horrible nailer." If the framer does not improve or change, then you may have to tell him that he is not suited for the work and should look for work elsewhere.

10. Pitch in and be a good example, especially if the job is one that nobody wants to do. You should not feel that any task is beneath you.

11. Be courteous. Everybody likes to hear "please" and "thank you." Saying "thank you" is a good way to finish up without giving the workers the sense that they are dismissed.

Respecting your framers will help keep them motivated, and help get the job done right.

Competition as a Motivator

It is sometimes possible to create competition that will provide enjoyment for your framers and increase productivity. Here is an example:

A while ago, I had a couple of hammers left over from tools I had purchased for a training class. On the job, we were framing a two-story hotel with two long walls on either side. I woke up one morning asking myself, "How can I make these walls go quicker?" I decided to create a competition by splitting the four framers into two teams, with one team on each side of the hotel. The winning team— the one that got their wall up first—would get the hammers. With the competition, the framers enjoyed the day and got a lot more wall framed than normal. Healthy competitions can help provide motivation.

Goals

One of the best management tools is goal-setting. It develops motivation by creating a reason to work productively, gives you a tool for judging the productivity of a framer, and provides a benchmark for discussing each framer's daily tasks. Goals should be set for different time periods, ranging from the entire length of the job, to daily or task goals. Goals can be written down, or you can go over them in a conversation with your framers.

Goals for the job are usually defined in the beginning by your schedule and manpower.

It helps to break down your overall project goals into goals for each part of the job, like the first-floor walls, the joists, and the rafters. Once you know the goals for the major parts of the job, you can begin to set your daily goals.

Set daily goals the first thing in the morning. You might want to think about them and who you will assign to each task on your way to work. After assigning the tasks, ask each individual to set their own goals for the day, which you can review with them.

Framers sometimes think they can get more work done in a day than they actually can. In this case, all you have to do is agree with their goals, and encourage your framers to achieve them. If, on the other hand, they set their goals at a lower rate of productivity than you expect, review their goals with them, and see if you can teach them faster ways to achieve them. You might do a little of their work for them so they can see how fast it is supposed to be done.

If you can't agree on a goal with a framer, give him another task, and assign his original task to someone else. At the end of the day you can compare how much work the other person accomplished with what you and the first framer expected, then determine which one of you was more on-target. This takes time and effort on your part, but sometimes that's what's needed to create motivation—which will save time in the long run.

It's important to review goals when your framers are done with their tasks—either at the end of the day or the next morning before you set new goals. This will show framers that goals are important. It

Delayed Communications

Organizing your communication is another important task of a lead framer. For example, you might realize that you need more material or hardware, but the person who orders it is not on the site. By the time you see that person, however, you are onto another task and may forget to let them know what you need.

The easiest way to avoid this problem is to carry a small notebook in your pocket, and write down what you need, whether it's information or materials. Get used to checking your notebook whenever you talk to the people who supply your material or process your change orders, or when you are making phone calls. The notebook acts as a memory aid when communications are delayed.

Assigning Framers Tasks

As lead framer, your most important job is to assign framers to tasks. If they are unfamiliar with the tasks, it's part of your job to teach them how to do the work. It may be tempting to just grab the right tool and take care of the problem yourself, but if you don't teach your crew, they won't be able to work independently, and neither will you.

Organizing your crew and assigning tasks can be the easiest part of your job, or it can be the most difficult. A lot of it has to do with the framers you have working for you, and the way you manage them. For example, one individual with a bad attitude can disrupt a whole crew, or a crew without proper direction can work all day and get little done.

When you first start leading, you'll quickly realize that it takes a lot of preparation to keep the whole crew busy all the time. As each framer finishes a task, you must have another task ready. If a task isn't ready, the framer(s) will have to wait around while you get it ready for them.

Dealing with Difficult Personalities

Chances are you will have to work with a difficult framer at some time. Common sense and creative thinking are good tools in these situations. Here are some difficult personality types:

- **The Complainer:** Looks for problems, not solutions. Suggestion: Tell him complaints are not helpful, he should look for a solution, and if he gets stumped, you're there to help.

- **The Back-Stabber:** Disrespectful and might wish to be lead framer. Suggestion: Confront disrespect or underhanded behavior openly. Explain to him how his actions affect the whole team. If you can't resolve the issues, speak to the framing or general contractor.

- **The Talker:** Probably a nice person, but as long as he keeps you talking, he isn't working. Suggestion: Be considerate, but after a short while, excuse yourself to get some work done.

- **The Perfectionist:** Wants every task done according to his standards and lets everyone know if something is less than perfect. Suggestion: Use this person's keen eye to uncover potential problems. Let him be a sounding board for new tasks.

In some cases, "difficult" framers may just have personalities that require relating to them a little differently. Try to use the individual strengths of each framer for the betterment of the team.

If you are working on a task and one of your crew needs something to do (and you don't have anything else for him to do), show him what you are doing so that he can help you or take over. Or have him get started on the next phase of the job. The point here is that if anyone is going to be standing around scratching his head, it should be you, because you can always use the time to plan for the next step.

Analysis of Crew Performance

For your framers to become better framers, they should have an understanding of how well (or not) they are performing their jobs. Crew analysis is the process of answering this question. It is important to know the capabilities of each framer, so you can assign him to the kind of task where he'll be most productive and know how much supervision or instruction he needs. Discuss these things with your framers. Their feedback will help you understand and evaluate them.

The "Framer Analysis" form can be used to evaluate your crew and to show your framers what aspects of their work are important to you. This is also a good format for deciding wage increases based on performance. The framer who consistently gets high ratings may get more money if he reaches a certain skill level.

To use the form, give the framer a rating from 1 to 10 for all the items listed. The "Value Factor" column in this form is an estimate of the comparative value of the productivity items. You can change these values to your own preferences. Enter your rating in the column titled "Framing Rating 1 to 10." Multiply the rating by the various value factors and put the results in the column labeled "Total." Add the total ratings. This will give you a value you can use to compare your framers' performance. You can use the Framer Analysis form for your own planning purposes or to show framers where they need to make improvements.

Framer Analysis

	Value Factor	×	Framer Rating 1 to 10	=	Total
Productivity					
1. Speed	7	×	_____	=	_____
2. Framing Knowledge	6	×	_____	=	_____
3. Framing Accuracy	5	×	_____	=	_____
4. Ability to use Framing Flow	4	×	_____	=	_____
5. Endurance	3	×	_____	=	_____
6. Consistency	2	×	_____	=	_____
7. Rate of Learning	1	×	_____	=	_____
Effect on Productivity of Other Framers					
1. Respect for Lead Framer	3	×	_____	=	_____
2. Cooperative	2	×	_____	=	_____
3. Positive Attitude	1	×	_____	=	_____
Effect on Efficiency of Company					
1. Attendance	4	×	_____	=	_____
2. Positive Attitude	3	×	_____	=	_____
3. Follows Safety Rules	2	×	_____	=	_____
4. Truck and Job Site Neatness	1	×	_____	=	_____
TOTALS					

Quality Control

In framing, the question of speed versus quality always comes up. You want to get the job done as fast as possible—but you must have a quality building, and quality takes time. The most important thing to consider is the structural integrity of the building. Once that requirement is satisfied, the faster the job can be done, the better.

It is a lot easier to talk about the importance of quality than it is to define it for a framer. Quality to one framer can be the product of a "wood butcher" to another framer. Framers learn under different lead framers who have different goals and objectives, and different standards of what quality workmanship is. You need to establish your own definition of quality of workmanship for the framers working for you.

The best way to do this is by observing or auditing their completed work, then giving them feedback on what you saw and what you would like to see. To audit the work, check a portion of what has been done. If that sample is done well, most likely the rest is done right. If you find a mistake, find out why it was made, correct any similar errors, and make sure the framer knows why this happened.

Audit Checklist

A checklist is helpful when you audit individual tasks. It will help you remember all the parts that need checking. For example, the following list could be used for shear walls.

Audit Guidelines

The following guidelines can be used to control the quality of experienced and new framers' work, and the work at the end of the job.

For an experienced framer you have worked with before:

1. Casually observe as part of routine.
2. Audit work after completion, or at regular intervals.

For new-to-the-task framers:

1. Review framing tips (at the end of this chapter).
2. When possible, demonstrate work.
3. Watch as the new framer gets started.
4. Ask the new framer to come and get you for review after the first piece is finished.
5. After a half hour to an hour, review the work.
6. End of day: review the work.
7. End of task: audit the work.

End of job:

1. Audit 10% of each individual task.
2. If mistakes are found, review all task work.
3. Correct all mistakes.
4. Check for omissions.

Framer-Friendly Tips

Whenever possible, allow the framer to fix their own mistakes. It's a great teaching tool.

Shear Walls Checklist

☐ Nailing pattern for sheathing

☐ Blocking, if required

☐ Distance between sheathing nails and the edge (³⁄₈" minimum)

☐ Nails are not driven too deep

☐ Lumber grade, if specified

☐ Hold-down sizes and location

☐ Hold-down bolt sizes

☐ Tightness of bolts

- ☐ Studs under beams
- ☐ Drywall backing
- ☐ Fireblocking
- ☐ Nailing sheathing
- ☐ Headers furred out
- ☐ Thresholds cut
- ☐ Crawl space access
- ☐ Attic access
- ☐ Dimensions of rough openings on doors and windows
- ☐ Check door openings for plumb
- ☐ Drop ceilings and soffits framed
- ☐ Stair handrail backing
- ☐ All temporary braces removed
- ☐ Joist hangers and timber connectors

Pick-Up Lists

Pick-up lists are important to keep things organized at the end of the job. The superintendent typically creates this list of tasks that have to be done before you are finished with the job. When the list is first given to you, review it to make sure everything is clear to you. It is sometimes easiest to ask the superintendent to accompany you around the site to make sure you understand exactly what he is trying to communicate.

You may have to consult the plans to get all the information you need in order to understand the work that needs to be done. If the superintendent does not have a written list, make your own list as you walk around the site discussing each task.

Remember that quality control is not just for the owner's benefit in the finished product. Quality control also makes your work go more smoothly. When your framers' cuts are square and true to length, the framing fits together a lot more easily.

If the building is square, when you cut joists and rafters, you can cut them all at once, the same length, instead of having to measure each one. When you get to the roof, the trusses will fit.

Organizing Tools and Materials

In addition to organizing and teaching the crew, you will have to organize your tools and materials. Each crew and job will require a different type of organization. To give you an idea of how to go about this, we will discuss three aspects: *tool organization, material storage,* and *material protection.*

Tool Organization

Following is an example of how the crew's tools might be organized using a job site tool truck.

Clear descriptions are important on pick-up lists. If you don't have a detailed pick-up list, you can count on returning to the job to fix at least one task.

General

Put tools away, in their designated place, after using them.

- Hang safety harnesses and lines on hooks.
- Stand sledge hammers and metal bars in corner.
- Place saws on saw table.
- Place nail guns in safety box.
- Place electrical tools in wood box.
- Hang up screwdrivers.
- Place metal wrenches and sockets in metal box.
- Place nails out of weather.
- Place trash in designated container.

Roll-up

Roll up largest, bulkiest items first.

- Four-way electric extension cords
- Air hoses
- Electric cords

Take equipment to truck in following order:

- Miscellaneous hand electrical tools
- Air hoses and electric cords
- Circular saws and old saw blades
- Air compressors (Drain every Friday.)
- Ladders

The person responsible for the truck:

- As soon as roll-up begins, start picking up and taking tools to the truck.
- Take tools from framers and put them in their place in the tool truck.

- Clean/organize truck when not busy putting tools away.
  — Put similar nails together.
  — Hang up rain gear.
  — Put tools in proper place.
  — Check and account for number of tools.
  — Put all loose garbage in bucket.

Nails

- Use up partial boxes of nails first.
- Follow established storage procedures. For example, starting at the right-hand side of back of truck
  — 1st: 16d sinkers   4th: joist hanger nails
  — 2nd: 8d sinkers   5th: concrete nails
  — 3rd: roofing nails   6th: fascia nails
- On right-hand side under seat, 10d gun nails.
- On left-hand side under seat, 8d gun nails.

Roll-out

- Check oil in air compressors every morning.
- Oil nail guns every morning.
- Check oil in circular saws the first of every month.
- Check staging and ladders.
- Check safety devices in all tools.

This list should be discussed at the first crew meeting on the job, then the list should be posted on the tool truck.

Material Storage

1. If your lumber is being dropped by a truck, check to make sure the lumber is loaded so that the items being used first are on top. You might need to contact (or have the superintendent contact) the lumber company to make sure they think about the loading order. Sometimes it helps to make up a quick list to help them out. For example:

   - Treated mudsill plate
   - Floor joists
   - Floor sheathing
   - Wall plates
   - Studs
   - Headers
   - Wall sheathing
   - Rafters
   - Roof sheathing

   The lumber company may not be able to load the material exactly the way you want, but a little concern for the loading order can make a big difference in the amount of lumber you have to move.

2. When using a forklift, store like items together so that you do not have to move other material to get at what you need.

3. When storing items, always think about where you are going to use them. If you don't have a forklift, store them as close to where you are going to use them as possible.

4. If you have to store items in front of each other, make sure the items needed first are available first.

5. Consider using carts or other mobile devices for moving lumber in the building.

Pallet jack and drywall cart for moving lumber

Material Protection

Material protection also requires you to consider accessibility and time. You can spend a lot of needless time moving and protecting material. You can also end up reducing the quality of your building by not taking care of your material. Consider the following:

1. If the specifications indicate a certain procedure for protecting your material (usually the case on larger jobs), then you need to follow them.

2. Use scrap lumber to keep your material out of the dirt.

3. If lumber is left in direct sun, the exposed sides will dry out more than the unexposed sides, and cause it to warp. The warp will make framing difficult, and walls curved.

4. Moisture loss or absorption from lumber causes shrinkage or swelling. If the shrinkage or swelling is uneven or happens too quickly, the wood fibers can break and cause the lumber to warp.

5. Fungal growth occurs when moisture content reaches 20%, and the air temperature is between 40 and 100 degrees. Fungal growth causes decay and stain.

6. The moisture content of green lumber is little affected by rainfall. But if it is not used right away, green lumber is more susceptible to fungal decay and stain.

7. Posts, beams, and timbers are always green. Seasoning checks will occur, but will not affect the structural performance. The more this lumber is protected, the less it will check, and the easier the installation will be.

8. Always cover lumber if you are expecting snow or other bad weather. It is easy to lift the cover to remove the snow when you need to use the lumber. On the other hand, if the lumber is being used up quickly and is not adversely affected by the environment, covering it may be a waste of time.

9. If lumber is delivered with covers already on, leave them in place as long as possible. If you are using only small quantities at a time, consider pulling the lumber out at the ends to leave the cover on. With engineered wood products such as glu-lam beams, you may be able to just uncover the ends for bearing.

10. If moisture absorption is expected on full bundles of sheathing, cut the banding to prevent edge damage due to expansion of the sheathing.

11. When lumber is covered, allow ventilation so that the sun does not create a greenhouse effect that will promote mold growth.

Teaching Framers

You have to take training seriously if you want your framers to take learning seriously. You are a teacher whether you want to be or not. The only question is whether you are a good teacher.

Good teachers have confidence in their knowledge and an understanding of those working for them. Remember:

- A picture is worth 10,000 words.

- A demonstration is worth 100 pictures.

- Tell your framers what you are going to tell them, tell them, then tell them what you told them. Repetition makes learning easier.

- It is important that framers understand the structural significance of their work.

Some of your teaching will apply to all of your crew. For example, special nailing may be specified for double wall plate joints for the whole building. A crew meeting is a good time to inform the whole crew all at once.

Hold a crew meeting before you start a job, then once a week after that. Monday morning meetings can help ease everyone back from the weekend. You can have these meetings right before or after your safety meetings, while you already have everyone together. It's nice if the crew meeting can be a relaxed time, while still covering important points such as:

- Task assignments
- Crew procedures (crew organization)
- Tool organization (tool truck)
- Job-specific items

Teaching While Assigning Tasks

Most of your teaching will occur when you are assigning tasks to framers. You won't have to say anything to your experienced framers, but new or apprentice framers benefit from seeing you follow a certain procedure to make sure you don't forget anything.

When assigning a task:

- Always assume your framers are seeing the task for the first time.
- Explain everything you know about the operation.
- As you're explaining the operation, tell your framers why it's done this way.
- Ask them if they understand. (Have them explain it to you.)
- If they ask you a question and you don't have the answer, tell them you'll find out and get back to them.

Check on them:

- After five to ten minutes.
- Repeatedly until you're confident that they know what they're doing.

When you're teaching a framer trainee, remember that they're learning as a student, so expect that it may take a little while for them to catch on. Don't expect all trainees to learn instantly, but always assume they want to learn.

The best way to communicate how to do a job is to actually do the job, and let the trainee watch. At the same time, explain as much of what you're doing as possible.

If you are showing an apprentice how to nail off plywood, use the following sequence:

1. Tell them the nailing pattern.
2. Ask them if they know what "nailing pattern" means.
3. Tell them about keeping the nail ⅜" away from the edge of the plywood.
4. Tell them to angle the nail slightly toward the edge of the plywood.
5. Tell them when they need to use a regulator or depth gage, and show them how.
6. Tell them how to avoid breaking the plywood surface.
7. Demonstrate use of a nail gun.
8. Watch them while they shoot a couple of nails in.
9. Ask them to come and get you to check their work after they have finished two sheets.
10. Check their work closely to make sure it's done properly.

Using this type of checklist will help you remember all the items that should be covered. This list also helps with assigning framing tasks.

Framing Tips for Every Task

When the lead framer is assigning tasks, he has to decide what information he has to tell the framer before starting the task. If the crew member has never done the task, the lead framer needs to explain it. If the framer has done this task many times, little needs to be said. If the framer's knowledge is not clear, it's best to review the task with him.

There are certain "tips" that experienced framers have developed for each task. Use the ones provided in this section or keep your own list to help your crew members.

Use your legs to support studs while spreading.

Building Wall Tips
Material Movement for Walls

1. Locate wall framing so that once the wall is built, it can be raised into position as close to where it finally goes as possible.

2. Spread the headers, trimmers, cripples, and sills as close to their final position as possible.

3. Eight is an average number of 2 × 4 studs to carry.

4. You can use your leg to stabilize the studs you are spreading. Stabilize them with one arm and one leg to free up your other arm so that you can spread them one at a time. This way you won't have to set them down, then pick them back up to spread them. (See photo.)

5. Select a straight plate for the top and double plates, and position any crown in the double plate in the opposite direction of the top plate crown. This will help straighten out the wall.

Nailing Walls

1. Nail the headers to the studs first. Make sure that they are flush on top and on the ends of the headers.

2. Nail the trimmers to the studs. Make sure that they are up tight against the bottom of the header and flush with the sides of the stud.

3. Nail the studs and cripples to the plates. Nail sills to the cripples and the trimmers. Make sure that all the connections are tight and flush.

Framing Tips *(continued)*

Teaching wall-building

Squaring Walls

1. Align the bottom plate so that when it is raised, it will be as close to the final position as possible.

2. Attach the bottom plate to the floor along the inside chalk line for the wall. Toenail through the bottom plate into the floor so that the sheathing won't cover the nails. If the wall is in position, it can be nailed on the inside, and the nails can be pulled out after the wall is raised.

3. Use your tape measure to check the diagonal lengths of the wall.

4. Move the top part of the wall until the diagonal lengths are equal. Example: If the diagonal measurements are different by one inch, then move the long measure toward the short measure by one-half-inch diagonal measure. Make sure the measurements are exact.

5. Once the diagonals are the same, check by measuring the other diagonal.

6. Temporarily nail the top of the wall so that it will not move while you are sheathing it. Make sure you nail so that your nails won't be covered by the sheathing.

Teaching joisting

Joisting Tips

Material Movement for Joists

1. Material movement is a major part of installing joists.

2. Always carry the joists crown-up. This way, you can spread the joists in place, in the right direction, without having to look for the crown a second time. It's easier to look for the crown on the lumber pile than when it is on the wall.

3. Check on the size of joists and positions needed. Try to spread the joists on the top of the pile first so you won't have to restack them.

4. Check your carrying path for the joists. Sometimes you can reduce your overall time by making a simple ramp or laying a joist perpendicular to those already in place.

Cutting Joists to Length

1. Cut joists after spreading.

   - Spread joists on layout, and tight to rim joists.
   - Chalk cut line.
   - Lift and cut each joist in sequence.

2. Cut joist on lumber stack.

   - Measure joist lengths.
   - Cut multiple joists on lumber pile.

Nailing Joists

1. Position joist on layout and plumb.

2. Nail through rim joist into joists, making sure joist is plumb.

3. Toenail through joist into double plate. Nail away from end of joist to prevent splitting.

Framing Tips (continued)

Rafter Tips

Cutting Rafters

1. Figure cut lines for rafters, and check measurements before cutting.
2. Install common rafters first.
3. Cut three rafters.
4. Check two to see if they fit. If they fit, leave them in place and use the third as a pattern for remaining cuts. If they don't fit, cut to fit or save for hip or valley jacks.
5. Cut balance of common rafters and install.

Installing Ridge Board

1. Figure height for ridge board.
2. Install temporary supports for the ridge board.
3. Install ridge board.

Nailing Rafters

1. Toenail common rafters on layout into double plate.
2. Nail on layout through ridge board into rafter.

3. Cut hip and valley rafters.
4. Cut jack rafters.
5. Set and nail hip or valley rafter.
6. String line centerline of hip or valley.
7. Layout hip or valley rafter.
8. Toenail jack rafter on layout through rafter into double plate.
9. Nail jack rafter to hip or valley rafter.

Sheathing Tips

Floor Sheathing

1. Make sure the first piece goes on square.
2. Chalk a line using a reference line and the longest part of the building possible.
3. Align the short edge of the plywood with interior joists, the long edge with the rim.
4. Pull the layout from secured interior joists.
5. Nail the plywood to align with the chalk line and layout marks.

Teaching rafters

Wall Sheathing After Walls Are Standing

1. Make sure the first piece goes up plumb. If you are installing more than three pieces in a row, use a level to set the first piece plumb.

2. It is easier to install the plywood if you are able to fit a 16d nail between the concrete foundation and the mudsill.
   a. Place two nails under each piece near each end.
   b. Remove the nails when you are finished.

3. The easiest and fastest way to handle an opening in the wall is to just sheath over it, then come back and use a panel pilot router bit to cut out the sheathing.

Roof Sheathing

1. Make sure the first piece goes on square.

2. Chalk a line from one end of the roof to the other.
   - When measuring for the chalk line, make sure you consider how the plywood intersects with the fascia. The plywood may cover the fascia, or the fascia may hide it.

3. If the sheathing overhang is exposed, the sheathing could take a special finish.
   - If the exposed sheathing is more expensive than the unexposed sheathing, then often the exposed sheathing is cut to fit only the exposed area. In this situation, cut the sheathing so that it breaks in the middle of the truss or rafter blocking.

4. 24" is the minimum width of any row of sheathing. Check before you get to the last row in case you need to cut a row so the last row will be at least 24".

Nailing Sheathing

1. Read the information on the stamp on each piece of plywood. Make sure you are using the right grade. Sometimes the stamp will tell which side should be up.

2. There should be at least a ⅛" gap between sheets for expansion.

3. The heads of the nails must be at least ⅜" from the edge of the sheathing.

4. Make sure that the nail head does not go so deep that it breaks the top veneer of the sheathing. Control nail gun pressure with a pressure gage or depth gage.

5. Angle the nail slightly so that it won't miss the joist, stud, or rafter.

6. Use the building code pattern for walls, floors, and roofs. Always check the plans for special nailing patterns. (Most shear walls have special patterns.)

Teaching sheathing

Framing Tips *(continued)*

Installing Hold-Downs (while walls are being built)

1. Locate wall hold-downs on plans and check details.
2. Locate holes to be drilled for hold-downs, anchor bolts, and through-bolts.
   - Measure location for through-bolts.
   - Center hold-downs on plates.
   - Center hold-downs in post or align with anchor bolts.
3. Drill holes.
4. Nail post into wall.
5. Nail sheathing to wall.
6. After wall is standing, install hold-downs, bolts, washers, nuts, and through-bolts.
7. Tighten all bolts and nuts.

Installing Hold-Downs (after walls are built)

1. Select a work area (large, close to material).
2. Check the plans for the location, quantity, and other details of hold-downs.
3. Collect all material and tools needed.
4. Spread hold-down posts for common drilling (Cut if necessary.)
5. Mark hold-down posts for drilling.
6. Drill for posts with holes $\frac{1}{16}$" larger than the bolts.
7. Loosely attach hold-downs, bolts, washers, and nuts to posts.
8. Spread hold-downs to installation location.
9. Drill holes for through-bolts if necessary.
10. Place hold-down in wall.
11. Place through-bolts into hold-downs where required.
12. Tighten all nuts.
13. Nail posts to plates.
14. Nail sheathing to posts.

Teaching installing hold-downs after walls are built

Framing Tips (continued)

Framer-Friendly Tips

It is best to install hold-downs as walls are built so you don't have to worry about plumbers and electricians getting their pipes and wires in your way.

Teaching installing hold-downs while walls are being built

Removing Temporary Braces

1. Remove temporary braces only after the walls have been secured so that they will not move.

2. A sledge hammer provides a fast and easy way to remove the braces.

3. Knock a number of the braces off at one time. Be careful that no one steps on the nails before you remove them.

4. Put the removed braces together.

5. Hit the point end of the nail to expose the nail head.

6. Use a crowbar to remove the nails.

7. If you do not have many braces, a hammer is an easy way to remove them.

Teaching removing braces

Framing Tips (continued)

Material Organization and Cutting

Material Organization, Mitre Saw

1. Set up your table in convenient locations for moving lumber in and out.

2. Place the incoming material as close as possible to the side of the saw where you will be positioning it to cut.

3. As it is cut, stack the lumber in a pile that is neat and easy to pick up and carry or lift with a forklift.

4. Put scrap wood that you will be cutting into stacks nearby, maybe under the saw table.

Cutting with a Mitre Saw

1. Set your length gage for multiple cuts.

2. Cut your first piece, then check the cut for square (both vertical and horizontal) and correct length.

3. Check the second and tenth piece for square and length.

4. Check every tenth piece after that for length.

5. Keep lumber tight against lumber guides, but don't bang them so that they move.

6. Remove any sawdust near the guides.

7. Respect the saw! If you don't, there is a good chance you will hurt yourself or your fellow framers.

Fractions

Many apprentice framers are not familiar with fractions, and some might be embarrassed to admit this. It doesn't take long to teach fractions. Ask the framer to show you where $^{11}/_{16}$" is on the tape. If he can't do it easily, draw a duplicate of a tape showing the different length lines. Mark the fractions on each line, and tell him to take it home and memorize it. Review as frequently as required to develop proficiency.

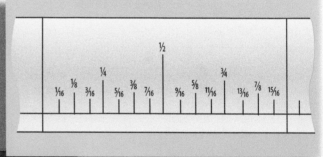

Reading a tape measure

Teaching mitre saw

Planning and Scheduling

While lead framers are not responsible for developing project costs or schedules, they are asked for input into the decisions of others who must estimate and schedule construction. The superintendent, for example, might need to know if he can meet a deadline with the crew that is in place; a lead carpenter might need to know if there is enough material available to complete the job; or the framing contractor might need to know if any labor can be spared to send to another job.

To answer these questions, the lead carpenter must understand and appreciate the importance of the construction schedule and budget. This means thinking ahead and looking at the project as a whole, while also focusing on the details. It means evaluating the crew's ability to perform its job at a particular time under a given set of conditions.

Using the Crew Effectively

A crew includes both labor and equipment required to install materials. On any given day, the makeup of the crew can change. One person may not show up or may be sent to another job. At other times you may have to absorb extra manpower on short notice. Equipment you expected to have available may not be there or, with little notice, you may have the benefit of equipment. When faced with these situations, a seasoned lead framer draws on his or her experience and makes the necessary adjustments to either *push* the job or, at the very least, maintain the momentum.

Staying on Schedule

The lead framer should understand the labor hours required for each task, and be able to know in his or her mind if the schedule is realistic and can be maintained. Perhaps more important is recognizing potential problems before they become real problems. In a matter of hours, what seems to be a minor glitch can become devastating to the schedule. For example, if fuel has not been requested for a piece of equipment, production may be forced to stop while waiting for a fuel delivery.

Never underestimate the importance of realistically measuring the crew's ability to perform the work. Keep in mind that most jobs, unless very short in duration, are scheduled well before the work begins, and job durations are usually based on optimistic job site conditions. During the course of construction, a monthly schedule is broken down to weekly schedules, and weekly schedules are broken down to daily schedules.

Before committing the lead framer to a schedule, the framing contractor normally has agreed to the means and methods that will be followed for the job. If the estimator has based the estimate on a crew that performs differently than an average crew, the schedule may or may not have allowed enough time for the work.

Framer-Friendly Tips

If you're working for a contractor, once you start managing a crew, your own productivity will be measured by theirs.

Planning for Materials

Avoiding Slow-Downs

Once a productive and effective crew has been assembled, nothing can slow that crew down faster than a shortage of material. Construction project estimators and schedulers often look at past project costs for material, labor, and equipment needs and costs. They may make adjustments to these figures based on input from the field, allowing for factors such as a more experienced work force, or new equipment that will make the work go faster. Nevertheless, material shortages can still occur. The lead framer needs to keep an eye on the rate at which materials are being used, and communicate material needs to the superintendent.

Taking Waste into Account

Project estimators perform quantity takeoffs that are really a best guess of how much material will be needed for the job. Waste is a concern in the quantity takeoff for any area of construction. There is some inevitable waste in framing lumber, depending on spans, wall heights, and the grade of lumber. A rule of thumb for lumber waste is 5%–10%, depending on material quality and the complexity of the framing.

Making Sure You Have the Correct Stock

The lead framer should be made aware of any material lists, structural framing drawings, shop drawings, engineered drawings, or cut lists that have been prepared for a framing project. This information is critical to ensure that the correct stock (lengths and widths) is used in the assembly of the frame. Read all notes on the drawings and find out whether the plans being used are the most recently amended or approved version.

Using the plans and shop drawings, the lead framer can determine which material to use for cripples, jacks, headers, blocking, and other miscellaneous members.

"Short" or "Will Call" Deliveries

Keep in mind that many initial stock deliveries are "short," meaning that as the project nears completion, someone is responsible for ordering just enough materials to complete the frame. This is sometimes referred to as "will call." The lead framer needs to know in advance if this strategy is being used.

On some projects where material storage and handling are restricted, a "just-in-time" delivery schedule may be necessary. This means that the lead framer must, in some cases, anticipate material and equipment needs on a daily basis.

In "will call" or "just-in-time" situations, the lead framer must be made aware of any problems in deliveries and must estimate and plan material use in order to maximize the productivity.

How Change Orders Affect the Schedule

A lead framer may be given instructions to perform change orders with little regard for how the change will affect time, cost, and crew productivity.

The time and cost of change order work varies according to how much of the installation has already been completed. Once workers have the project in their mind, even if they have not started, it can be difficult to re-focus. The lead framer may spend more time than usual understanding and explaining the change. Modifications to work in-place, such as trimming and refitting, usually take more time than was initially estimated. Post-installation changes generally involve some demolition. The change may come after finishes and trim are installed and may require protection of in-place work.

When faced with a change or a rework situation, the lead framer must break down the typical day into segments and estimate the impact on each segment. Say a change involves reframing an opening or creating a new opening in a wall that has been completed. The estimated time for the change should account for demolition, possible salvage of original materials to be reused, procurement of new materials required, and possibly a reluctance of the crew to perform the change. The time spent on the change will generally add time to work in progress. If the lead framer anticipated four openings per day and now has to reframe two, productivity for framing openings may drop to three per day until the change is complete. This will delay setting windows or installing exterior sheathing and other tasks. (See "Changes to the Plans" and "Extra Work" in the next section.)

Recordkeeping

Recordkeeping is quite possibly one of those tasks that you thought you were getting away from when you started framing. The reality is that recordkeeping is an important, but not necessarily major, task for the lead framer. There are three things you will want to keep records for: timekeeping, changes to the plans, and extra work.

Timekeeping

Timekeeping is easy, but you have to record it every day. If you don't, it's easy to forget and make a mistake that is not caught until the payroll checks come out. Most companies provide forms that can be filled out at the end of every day. You will need some type of an organizer to store your time cards and other records. For a small job, an aluminum forms folder works well. These folders are durable and keep the rain out. If you are working on a big job, you will probably need something like a builder's attaché to keep all your papers organized. Your time cards can be kept in your organizer so you always know where they are.

Changes to the Plans

Changes to the plans should always be recorded when they occur. Changes may be conveyed in conversation or in writing. Because the time when you receive the changes is not always the time you will be doing the work, it is important to record the information so that you will not forget it. The best place to record changes is on your plans. Write it on the sheet where you will see it, then write the date and the name of the individual who gave you the change. If it was given to you on paper, keep that document in your organizer after you have written the change on the plans. You can also tape the change to the plans. If there is not enough room to record the changes on the appropriate sheet, tape the information on the back of the prior sheet so you will see it when you are reading the sheet involving the changes.

Keeping your papers organized

Extra Work

The third recordkeeping task is recording change orders. This is important because if work is done that wasn't originally figured in the framing bid or contract, it must be documented in order to obtain payment. This can be a sensitive issue. Many times there is controversy over payment for tasks that are not clearly defined in the bid or contract. If at any time you are asked to perform work that you consider a change order, you should inform the person asking you to do the work right away that this extra work constitutes a change order, and that you expect to be paid for it. The person requesting the extra work can then decide whether they still want to make the change, knowing the extra cost it involves.

When you actually perform the change order work, make sure you record the work done and the cost to be billed. If you are to be paid on a time and material basis, you need to keep accurate time records showing the hour of the day, and the date the work was performed.

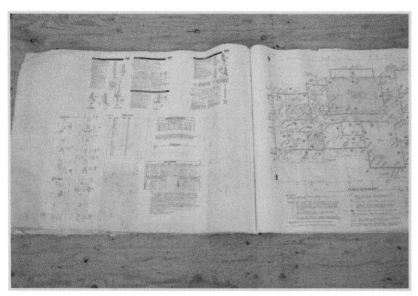

Plans with taped changes

Productivity

Your objective is most likely to frame a quality building for the least possible cost. Wages represent the greatest percentage of the total cost, and since wages are paid for time spent working, it follows that if you lessen the time it takes to complete a building, you thereby reduce the wages paid and, thus, lower the total cost.

There are various tasks involved in framing. Some of the tasks can be done as fast, or nearly as fast, by the least experienced framer as by the most experienced framer—for example, carrying and spreading studs. Suppose that a $10/hour person can do the same job as a $20/hour person, only 20% slower. You save $8/hour by having the less expensive person do the job.

Be careful, however, because if an inexperienced framer takes on a job beyond his capabilities, the chances for mistakes are great. This could result in wages paid for work incorrectly done, and higher wages paid for an experienced framer to find and correct the mistake.

It is ideal for you to have a balanced crew of experienced and inexperienced framers. This allows you the flexibility to fit the framer to the task. It takes planning to coordinate framing tasks so that they are done as inexpensively as possible, but it is time well spent for the money you will save.

The following are job titles and responsibilities for a typical large framing crew, divided into four categories: overall organization, walls, floors, and roofs.

Organizing

1. Lead framer

☐ Coordinate work schedule with framing or general contractor.
☐ Locate lumber drop.
☐ Make sure tools and supplies are available.
☐ Train framers.
☐ Solve problems.

2. Cut and chalk framer

☐ Cut stairs and rafters.
☐ Chalk lines.
☐ Assign framers to tasks.

3. Layout framer

☐ Layout for walls, joists, rafters, and trusses.

Walls

1. Plumb and line framer
☐ Plumb and line walls.

2. Wall spreader

☐ Spread all parts of a wall (studs, trimmers, cripples, headers, etc.) so they are ready to be nailed together.

Typical large framing contractor crew

3. Wall nailer

☐ Nail walls together.

4. Makeup framer

☐ Nail together a quantity of stud-trimmers, corners, backers, headers, etc., before they are carried to the walls.

Floors

Joisting

1. Rim joister

☐ Nail joists in position.
☐ Header out joists.

2. Joist nailer

☐ Nail joists in position.
☐ Nail on joist hangers.
☐ Nail blocking and drywall backing.

3. Joist spreader

☐ Carry joists into place.

Sheathing

1. Sheathing setter

☐ Place and set sheathing.

2. Sheathing nailer

☐ Glue joists ahead of setter and nail sheathing behind setter.

3. Sheathing carrier

☐ Carry sheathing to setter.

Roofs

1. Spreaders

☐ Spread and install trusses and rafters.
☐ Install fascia.

2. Blockers

☐ Install blocking and drywall backing.
☐ Set sheathing.

3. Sheathing packers

☐ Carry sheathing to setters.

Typical large framing contractor crew (continued)

Multiple-Framer Tasks

Some tasks require more than one framer, for example, lifting large walls. Any time framers have to be called from other tasks to perform a common task, care must be taken not to waste time. The cost per minute of a 5-framer task is 5 times that of a single framer task. The person organizing the task needs to take responsibility for making sure the task is ready for all framers. If it is lifting a wall, the organizing framer should make sure the wall is completely ready so framers don't stand around while one person makes last-minute adjustments.

Learning Curve

Studies have been done that show that as output is doubled, the time required decreases according to a constant ratio. The common ratio is about 4 to 5, or 20%. For example, the fourth set of stairs built will take 20% less time than the second.

Multiple Cutting Analysis

Multiple cutting becomes efficient when you have to cut a number of pieces of lumber the same size. Trimmers, cripples, and blocks are good examples.

To multiple cut, first spread all the lumber to be cut out on the floor or a table. Then measure each piece, mark each piece with your square, and finally cut each piece.

Analysis has shown it takes 36% less time to cut ten pieces of 2 × 4 when the tasks of spreading, then measuring, then marking, and then cutting are done for all the pieces at one time.

For very large numbers of cuts it may be worthwhile to make a template or a measuring/cutting jig.

Motion Analysis

Question: Can significant time (and money) be saved by moving faster?

Answer: Motion analysis studies have shown that something as simple as walking more quickly around the site can substantially raise productivity. For instance, if a framer spends 2 hours in an 8-hour day walking from point to point on a job site, a quick walk can save about 30 minutes per day over a relaxed walk.

Speed Versus Quality

Speed or quality: Which should it be? How good must the work be if it must be done as fast as possible? The two variables to consider when answering these questions are:

- Strength
- Attractiveness

First and most important is the structural integrity of the building. The second is creating a finished frame that will be pleasing to the eye. Once requirements for strength and attractiveness are satisfied, the faster the job can be done, the better.

Material Movement

Framing requires a lot of material movement. It is estimated that one-quarter to one-third of a framer's time is spent moving material, so any time or energy saved is a cost reduction.

The following hints will help you save time, energy, motion and, in the last analysis, money.

- Whenever material is lifted or moved, it takes time and energy; therefore, move material as little as possible.

- When stacking lumber, consider the following:
  - —Where will it be used next?
  - —Will it be close to where it is going to be used?
  - —Will it be in the way of another operation?
  - —Will it obstruct a pathway?
- Always stack material neatly. This helps to keep the lumber straight and makes it easy for framers to pick up and carry it. Stack 2 × 4 studs in piles of eight for a convenient armful.
- Have second-floor lumber dumped close to the building so framers can stand on the lumber stack and throw it onto the second floor.
- When stacking lumber on a deck, place it where walls will not be built, so it will not have to be moved again.

- Use mechanical aids, such as levers, for lifting. Remember your physics—the longer the handle in relation to the lifting arm, the easier it will be to lift the load.
- Two trips to the lumber stack or tool truck cost twice as much as one trip. If you have to go to the tool truck for a tool, check to see if you need nails or anything else.

Framer-Friendly Tips

Pieces of pipe come in handy for rolling stacks of lumber. On bigger jobs with forklifts, use pallets for collecting and moving material, such as multiple blocks.

Tool Maintenance Schedule

Draw up a schedule such as the one shown in the following table for your specific equipment; post it, and assign a reliable crew member to take charge of it.

Equipment	Schedule	Maintenance Operation	Lubrication
Worm-drive saw	First of every month	Check oil	Heavy-duty saw lubricant
Nail guns	Each day before using	Oil	Gun oil
Compressors	Before using every day	Check oil	30-weight non-detergent
	Every Friday	Empty air and drain tanks	
	First of each month	Check air-intake filter	
	Every month	Change oil	30-weight non-detergent
Electric cords	First of each month	Test cords color code	

Framing Technology

Technology is never going to do away with framing, but it will make it easier. You may not come into contact with much technology, but it is good to have an understanding of the possibilities. Many suppliers can provide walls and sections of floors and roofs already framed ready to be shipped and installed. In these situations you have to deal with the weight of setting them in place and then the corrections to make sure everything fits. If you have the right equipment (boom truck, crane, forklift, etc.) and accurate planning, it can be a very efficient way to frame.

Some suppliers are also supplying construction details complete with a notebook for organization and laminated jobsite plans to go with the details. All the material can be cut to length. On these jobs staying organized is the biggest job. If you are not organized you will spend all your time looking for parts.

On bigger framing jobs technology is providing the organization. New software now makes it easy to keep track of RFI's (Request for Information) and communication between the lead framer, the superintendent, the construction office, the architect, and the engineer.

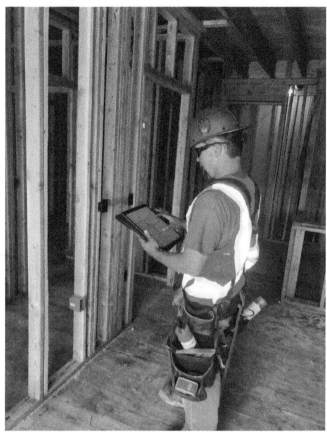

Lead Framer Studying Plans on Tablet

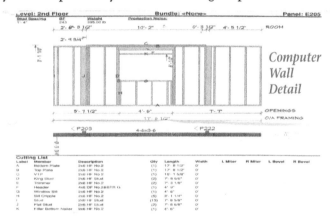

Computer Wall Detail

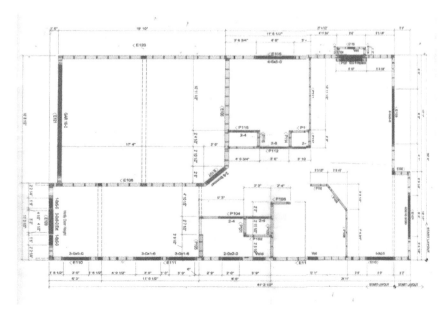

Laminated Computer-Generated Jobsite Plans Matched to Details

Conclusion

When you started reading this chapter, you were probably hoping for some nice clean answers on how to manage a crew—answers that you could put to use tomorrow. Now you are probably thinking that you have more questions than you did when you started reading—and that's the way it should be. Managing a crew is a never-ending job that will challenge you every day. The information presented in this chapter should give you a base for the common-sense decisions you will have to continually make in response to the questions that come up as you manage your crew.

Framing Crew Start 1979

A career as a Framing Contractor

Office

Framing Crew Finish 2013

Chapter Fifteen
SAFETY

Contents

Chapter Fifteen

SAFETY

If you have been framing long enough to understand advanced techniques or to be considering a career as a lead framer, you have probably seen enough accidents to make you aware of the importance of safety. Common sense will help guide you in knowing what is safe and what is not, but you must also be aware of the potential dangers. This information is usually acquired from the lead framer who taught you, from apprenticeship classes, and from weekly safety meetings, as well as state and federal regulations for the job site.

The safety topics presented in this chapter are not intended to be a complete list, but rather to cover the items you will come in contact with or have questions about most often.

Personal Protective Equipment

What we wear can either help prevent accidents or help cause them. Think about what you are going to do during the day, and prepare for it. It's a good idea to discuss personal protective equipment needed for specific tasks at your safety meetings. Keep an eye on new framers so you can detect any potential safety problems.

Hard hats are the symbol of the construction industry. Some jobs require that hard hats be worn. The Occupational Safety and Health Administration (OSHA) says that a hard hat needs to be worn if there is a possible danger of head injury from impact, or from falling or flying objects.

Eye protection is required by OSHA when there is a reasonable probability of preventable injury when equipment such as a nail gun is used. Eye protection can be provided by safety glasses. Safety glasses can be found that are lightweight and also look good. They should always be worn when using power saws.

Safety glasses

Ear protection is recommended when you are exposed to high levels of noise. High noise levels can cause hearing impairment and hearing loss, as well as physical and psychological stress. There is no cure for hearing loss caused by exposure to noise. Framers are exposed to these high levels at various times, not so much from their own work as from surrounding operations. The easiest way to protect yourself from hearing impairment is to keep disposable earplugs handy. They are easy to use and once they are in, you barely notice them.

Foot protection can be provided by a pair of leather work shoes or boots with hard soles. The boots will help protect your ankles. Steel toes provide extra protection for your toes and can be useful as support for lumber you are cutting. Rubber boots are good in wet weather and provide an extra measure to prevent electric shock.

Pants and shirts should be fit for work. If they are too loose, there is the chance they can get caught in something like a saw or a drill and pull you into the drill bit, which might throw you off a ladder. If your pants are too loose or frayed at the bottoms, they can cause you to trip and fall. Be careful with other clothing, such as belts and coats, so that they don't hang loose and get caught.

"Oh, my aching back." Everybody has heard those words. In fact, back injuries are the most common type of injury in the workplace. Framing is lifting-intensive work—so measures to prevent back injuries deserve your attention. Stretching each morning and strengthening exercises are good for your back, but more important is making sure you lift properly. (See "Proper Lifting" photo.) Make it a point to use your legs to lift, and not your back. When you are lifting walls, remind your crew to lift with their legs. When picking something up, bend your knees and keep your back straight. When carrying, keep objects close to your body, and avoid twisting and jerky motions.

Hand Tools

Nail guns are one of a framer's most commonly used tools. They are also one of the most dangerous. Most framers can show you a scar from having shot themselves with a nail gun. Fortunately, many of these injuries are not serious. However, there have been instances where serious injury or death has occurred. Following are some very basic guidelines that will help you operate a nail gun safely. (Always familiarize yourself with the manufacturer's complete operating instructions.)

- Wear safety glasses.

- Do not hold the trigger down unless you're nailing.

- Be careful when nailing close to the edge. The push lever at the nose of the gun can catch the wood and allow the gun to fire without the nail hitting the wood, allowing the nail to fly toward whatever is in line with the gun.

- Always keep your hand far enough away from the nose of the nail gun so that if the nail hits a knot or obstruction and bends, it will not hit your hand.

- Never point a nail gun at anyone.

- Disconnect the air hose before working on the gun.

- Use a gun hanger when working at heights, or secure your air hose so the gun does not get dragged off or fall. (See "Nail Gun and Hanger" photo.)

- When nailing off the roof or high floor sheathing, move in a forward, not a backward direction to prevent backing off the edge.

- Move from top to bottom on wall sheathing so you can use the weight of the gun to your advantage.

Trainees are the most vulnerable to nail gun accidents. Make sure that when you are training new recruits on nailing with a nail gun, you instruct them on nail gun safety and the potential for accidents.

Circular saws have cut off many fingers. A healthy respect for them is the first step toward safety. Follow these basic guidelines (and the manufacturer's operating instructions):

- Wear safety glasses when operating a circular saw.

- Always keep your fingers away from where the blade is going.

- Never remove or pin back the guard on the saw. The saw guard has a tendency to catch on many cuts, especially angle cuts, which makes it tempting to pin the guard back. Aside from the fact that it is an OSHA violation, a saw can become bound in a piece of wood, and "kick back." If the guard is pinned back, this can result in serious injury such as cuts to the thigh.

Proper lifting

- Never use a dull blade. It will cause you to put excess directional force on the saw, which could cause it to go where you don't want it to.

- Disconnect from power if you are working on the saw.

- When you are cutting lumber, make sure that one end can fall free so that the blade does not bind and kick back.

As the teeth of a circular saw speed around at almost 140 miles per hour, it becomes very dangerous if not used properly.

Miscellaneous hand tools also need to be used properly for safety. The following guidelines apply to many hand tools:

- Make sure all safety guards are in place.

- Keep your finger off the trigger of power tools when you are carrying them to prevent accidental starting.

Nail gun and hanger

- Keep tools properly sharpened.

- Store tools in the locations provided.

- Before working on power tools, unplug them or take out the battery.

- Replace worn or broken tools immediately.

- Never leave tools in paths where they can become a tripping hazard.

To use a **powder-actuated tool**, you need to be trained by a certified trainer. Following are some of the basics that you will learn:

- You must wear safety glasses.

- Hard hats and hearing protection are recommended.

- Never point a powder-actuated nail gun at anyone.

- Before you fire, make sure no one is on the other side of the material you are firing into.

- Do not load the firing cartridge until you are ready to use it.

- If there is a misfire, hold the tool against the work surface for at least 30 seconds; then try firing again. If the tool misfires a second time, hold it against the work surface again for 30 seconds; then remove the cartridge and inspect the gun. Soak the misfired cartridges in water in a safe location.

- Powder-actuated tools need to be placed firmly against the work, perpendicular to the work to avoid ricochet.

It's also a good idea to say "fire" just before you pull the trigger, so the shot noise will not startle the workers around you.

Ladder extension

Ladders

Inappropriate use of ladders is the number one cause of falls. Ladders are used so often in framing that it is easy to overlook basic safety guidelines. Always remember the following:

- The feet of the ladder need to be on a stable surface so the ladder will be level.

- When ladders are used to access an upper surface, make sure they extend at least three feet above the upper surface, and secure the top to prevent them from being knocked over. (See "Ladder Extension" photo.)

- Do not use the top or top step of a step ladder.

- For straight or extension ladders, remember the 4 to 1 rule. For every four feet of height the ladder extends, it needs to be placed one foot out at the base.

- Check the ladder for defective parts, and remove any oil or grease on the steps.

- Never leave tools on the top step of a ladder.

Use common sense. If you are not sure that a ladder is safe, don't use it.

Fall Protection

OSHA provides that for unprotected sides and edges on walking or working surfaces, or for leading edges of six feet or more, framers must be protected from falls by the use of a guardrail system, a safety net system, or a personal fall arrest system. For leading edge work, if it can be demonstrated that these systems are not feasible or they create a greater hazard, then a plan may be developed and implemented to meet certain OSHA requirements.

Most deaths in the construction industry happen as a result of falls. Falls also cause many of the injuries that occur on the job site, according to OSHA.

Because framing can place you a story or more off the ground, it's important to work safely and to become aware of fall protection equipment, as well as systems (such as guardrails, safety nets, and covers) that can help prevent falls.

The following sections cover safety tips on equipment and systems. Using these can help protect yourself and others on the job site from falls.

The most commonly used fall protection systems in framing are the *personal fall arrest system* and the *guardrail system*. The guardrail system works well for a large flat deck. The fall arrest is better suited for pitched roofs.

Guardrails

Metal guardrail supports can be nailed to the outside of walls to support 2 × 4 railings. However, it is more common to see railing made out of 2 × 4 for the support and railings. These railings can be nailed on before the walls are raised. Following are some OSHA regulations on guardrails:

- The top edge of the top rail must be at least 42" (plus or minus 3") above the working level.

- Mid-rails must be installed at a height midway between the top edge of the top rail and the working level.

Fall arrest system

- The top edge of the guardrail system must be able to withstand 200 pounds applied within two square feet in an outward or downward direction.

- The mid-rail must be able to withstand at least 150 pounds in an outward or downward direction.

- When access is provided in the guardrail system, a chain, gate, or removable guardrail sections should be placed across the opening when loading operations are not taking place.

Personal Fall Arrest Systems

These systems typically consist of a full body harness, a lanyard, a lifeline, and an anchor. (See "Fall Arrest System" photo.) Each of these parts is available in many different types. Some of the OSHA regulations for these systems are listed below.

- D-rings, snaphooks, and carabiners must have a minimum tensile strength of 5,000 pounds.

- Lanyards and vertical lifelines must have a minimum breaking strength of 5,000 pounds.

- Anchors must be capable of supporting at least 5,000 pounds per framer.

- The system must be rigged so that the framer cannot free-fall more than 6 feet.

- The attachment point for a body harness is to be located in the center of the wearer's back near the shoulder level or above the wearer's head.

If you have new framers who are not used to working with fall arrest systems, you will need to spend some time with them to help them become familiar and comfortable with this equipment.

Rough Terrain Forklift Safety

To operate a forklift, you need to be certified. To obtain certification, you need to be trained and actually operate a forklift. The points below are intended as a refresher for those who are already certified, and as an introduction to those intending to be certified.

- It is the weight of the forklift and the position of the tires that keep a forklift from turning over. There is an imaginary triangle between the front two tires and the space between the back two tires. This is called the *stability triangle*. The center of gravity for the forklift lies within this triangle. As the forks with weight extend out, the center of gravity moves. If the center of gravity goes outside the stability triangle, the forklift will tip. Getting the feel for the location of the center of gravity and the stability triangle is important to safe operation before you start working with a forklift. A good way to start is to lift a load of lumber and extend it out next to the ground until the back wheels start to come off the ground.

- The center of gravity is also changed when the forklift is on sloped ground. The situation is exaggerated greatly if there is a load on the forks and they are extended. (See "Forklift Center of Gravity" illustrations.)

- If you are using a forklift and it starts to tip over, stay in the seat; do not jump out of the forklift.

- Before you operate any machine, be sure you are familiar with all the controls.

- Before you operate the forklift, do an inspection. Walk around the forklift checking for anything that does not look right, such as leaking fluids. Then get in the cab, start the engine, and check the gages and other controls.

- Never leave the forklift while the engine is running.

- Know the forklift hand signals. (See "Forklift Hand Signals" illustration.)

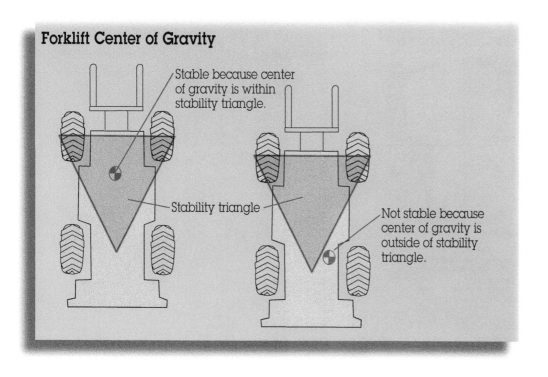

Forklift Center of Gravity

Stable because center of gravity is within stability triangle.

Stability triangle

Not stable because center of gravity is outside of stability triangle.

Forklift Center of Gravity

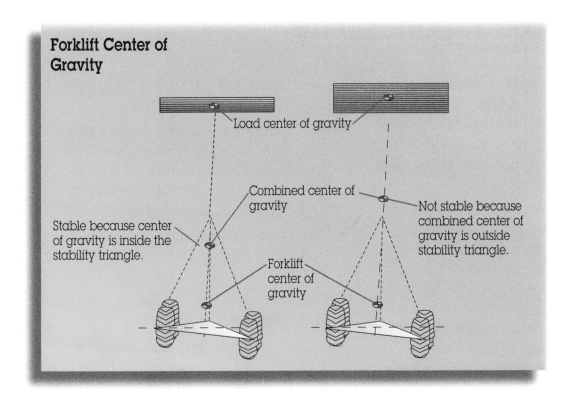

- Load center of gravity
- Combined center of gravity
- Stable because center of gravity is inside the stability triangle.
- Not stable because combined center of gravity is outside stability triangle.
- Forklift center of gravity

Forklift Hand Signals

FORKLIFT SIGNALS

1. STOP
2. RAISE LOAD
3. LOWER LOAD
4. TILT FORKS RIGHT
5. TILT FORKS LEFT
6. TILT FORKS UP
7. TILT FORKS DOWN
8. MOVE LOAD BACKWARD
9. MOVE LOAD FORWARD

INSTRUCTIONS TO SIGNAL MEN

1. Only one person to be signalman
2. Make sure the operator can see you and is able to acknowledge the signal given
3. Signalman must watch the load— the operator is watching you
4. Never raise or lower the load over other workmen, warn them to keep out of the way

WATCH FOR OVERHEAD LINES OR OTHER OBSTRUCTIONS

Source: Mason Contractor's Association of America

- Keep the forks close to the ground with or without a load.

- Always be looking for obstacles in your way like power lines overhead.

- Check the forklift load chart to make certain the forklift can handle the load you intend to move. The load capacity of a vehicle can be found on the identification plate and in the operator's manual. (See "Load Chart" photo.)

A forklift is a very big and heavy piece of machinery, capable of doing a great amount of damage. Make sure you treat it with respect.

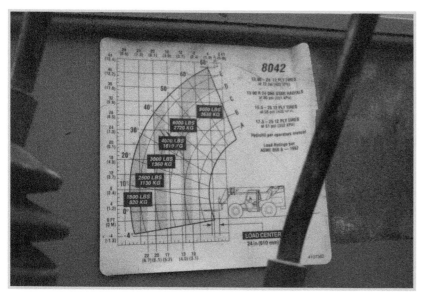

Load chart

Crane or Boom Truck Safety

When working with a crane or boom truck, the most important thing for a framer is to have good communication with the operator. Because you are typically out of audible range, you will need to use hand signals. You should use the industry-accepted standards shown in the "Crane and Boom Truck Hand Signals" illustration.

Housekeeping

Housekeeping is something we all grow up with. Some learn it better than others. On the job site, we all must practice good housekeeping because it affects safety and productivity. There are three main housekeeping issues: job site scraps, personnel debris, and tool organization.

Job site scraps are the cut-off ends of pieces of wood, lumber torn down that will not be used again, wrapping from lumber, empty nail boxes, and numerous other materials brought onto the job site that will not be used. You don't always want to take the time to attend to this debris at the moment it is made, but you do need to make sure that it is not left in a location that would pose a safety problem, such as in a walkway or at the bottom of a ladder. As you create the scraps, throw them in a scrap pile or at least in the direction of a scrap pile. Many cut-off pieces of lumber can be used for blocking and should be thrown in the direction of where they will be mass-cut later.

Whenever you have lumber with **nails** sticking out, pull the nails out if you are going to use the lumber again, or bend them if the lumber will be thrown away. It is easy to forget this, so be sure to attend to the nails while you are working with them and they are on your mind.

Personnel debris is the garbage individuals create personally, like lunch scraps and soda bottles. It makes it easier on your crew if you provide some sort of container near the lunch spot. You will be lucky if you don't have to keep reminding your crew that they are responsible for their own personal garbage.

Tool organization can save you time and a lot of aggravation. With tools that are in common use by the crew, it is important to return the tool to where it belongs. If tools are not stored in a location where other framers can expect them to be, a lot of time will be wasted looking for them. When rolling out your electric cords and air hoses, keep them organized. If you have to roll out in a walkway, try to stay to one side or the other. If you have to move your hose or cord, and it is rolled out underneath someone else's cord or hose, be careful that when you move yours, you don't drag theirs (and possibly their tools) along with you.

Housekeeping is sometimes difficult to organize, but getting it right can be a big asset for safety and productivity.

Mobile Crane and Boom Truck Hand Signals

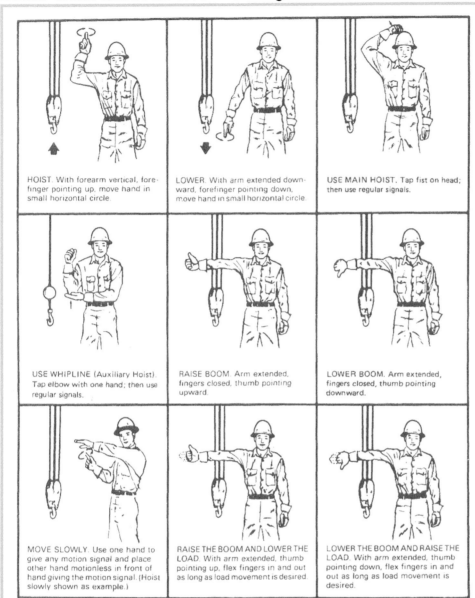

HOIST. With forearm vertical, fore-finger pointing up, move hand in small horizontal circle.

LOWER. With arm extended downward, forefinger pointing down, move hand in small horizontal circle.

USE MAIN HOIST. Tap fist on head; then use regular signals.

USE WHIPLINE (Auxiliary Hoist). Tap elbow with one hand; then use regular signals.

RAISE BOOM. Arm extended, fingers closed, thumb pointing upward.

LOWER BOOM. Arm extended, fingers closed, thumb pointing downward.

MOVE SLOWLY. Use one hand to give any motion signal and place other hand motionless in front of hand giving the motion signal. (Hoist slowly shown as example.)

RAISE THE BOOM AND LOWER THE LOAD. With arm extended, thumb pointing up, flex fingers in and out as long as load movement is desired.

LOWER THE BOOM AND RAISE THE LOAD. With arm extended, thumb pointing down, flex fingers in and out as long as load movement is desired.

Source: Reprinted from ASME B30.5-2004 by permission of the American Society of Mechanical Engineers, all rights reserved.

Framer-Friendly Tips

One framer should be appointed to direct the crane, and no one else should give directions.

Mobile Crane and Boom Truck Hand Signals *(continued)*

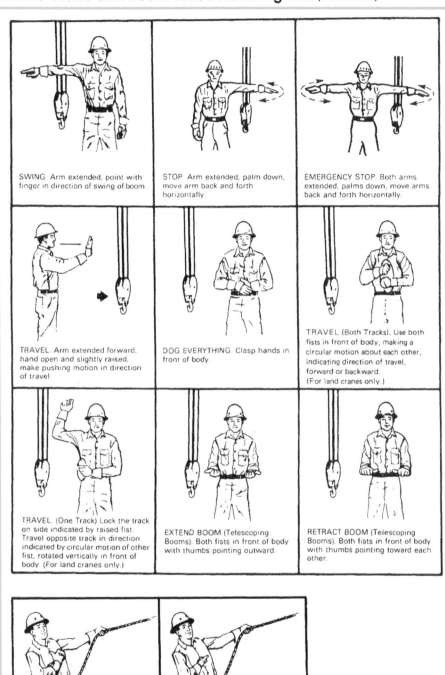

SWING. Arm extended, point with finger in direction of swing of boom.

STOP. Arm extended, palm down, move arm back and forth horizontally.

EMERGENCY STOP. Both arms extended, palms down, move arms back and forth horizontally.

TRAVEL. Arm extended forward, hand open and slightly raised, make pushing motion in direction of travel.

DOG EVERYTHING. Clasp hands in front of body.

TRAVEL (Both Tracks). Use both fists in front of body, making a circular motion about each other, indicating direction of travel, forward or backward. (For land cranes only.)

TRAVEL. (One Track) Lock the track on side indicated by raised fist. Travel opposite track in direction indicated by circular motion of other fist, rotated vertically in front of body. (For land cranes only.)

EXTEND BOOM (Telescoping Booms). Both fists in front of body with thumbs pointing outward.

RETRACT BOOM (Telescoping Booms). Both fists in front of body with thumbs pointing toward each other.

EXTEND BOOM (Telescoping Boom). One Hand Signal. One fist in front of chest with thumb tapping chest.

RETRACT BOOM (Telescoping Boom). One Hand Signal. One fist in front of chest, thumb pointing outward and heel of fist tapping chest.

Source: Reprinted from ASME B30.5-2004 by permission of the American Society of Mechanical Engineers, all rights reserved.

Safety Tips

These safety tips provide a quick summary of basic safety concerns.

1. Keep your work area neat underfoot, especially main pathways.

2. Keep work areas well lighted and adequately ventilated.

3. Don't leave nails sticking out; pull them or bend them over.

4. When lifting, always lift using your leg muscles and not your back. Always keep your back straight, your chin tucked in, and your stomach pulled in. Don't carry objects resting against your stomach; this will cause your spine to bend backward. Maintain the same posture when setting objects down.

5. When working on joists, trusses, or rafters, always watch each step to see that what you are stepping on is secure.

6. Never pin circular saw guards back.

7. Do not wear loose or torn clothing that can get caught in tools.

8. Wear heavy shoes or boots that help protect your feet against injury.

9. Wear hard hats where material or tools can fall on your head.

10. Wear safety glasses where there is any potential for injury to the eye.

11. Use hearing protection when operating loud machinery or when hammering in a small, enclosed space.

12. Wear a dust mask for protection from sawdust, insulation fibers, or the like.

13. Use a respirator whenever working in exposure to toxic fumes.

14. Assign a qualified person to regularly inspect the job site.

15. Make first aid supplies available and make sure they are replenished on a regular basis.

16. Post location and number of the closest medical facility.

17. Always secure ladders before using.

18. When exposed to a fall hazard, always wear a personal fall arrest system.

Conclusion

The safety topics in this chapter are just a small part of a total safety plan, but represent a good base. Remember, as a lead framer, you can't afford to forget about your own safety, not only for your own well-being, but because your framers will look to you to see how seriously you take it. It's easy to forget safety measures when you are in a rush on a job, but when you stop and think about it, the alternatives could be losing a finger or even risking death. It's not hard to see the importance of safety. With a little bit of effort, your jobs can be safe and productive.

Chapter Sixteen
HEALTHY FRAMING

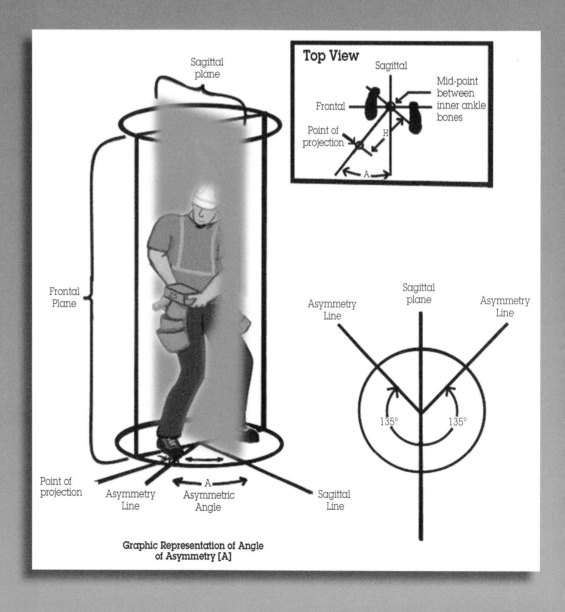

Top View

Sagittal plane

Sagittal

Frontal

Mid-point between inner ankle bones

Point of projection

H

A

Frontal Plane

Sagittal plane

Asymmetry Line

Asymmetry Line

135°   135°

Point of projection

Asymmetry Line

A
Asymmetric Angle

Sagittal Line

Graphic Representation of Angle of Asymmetry [A]

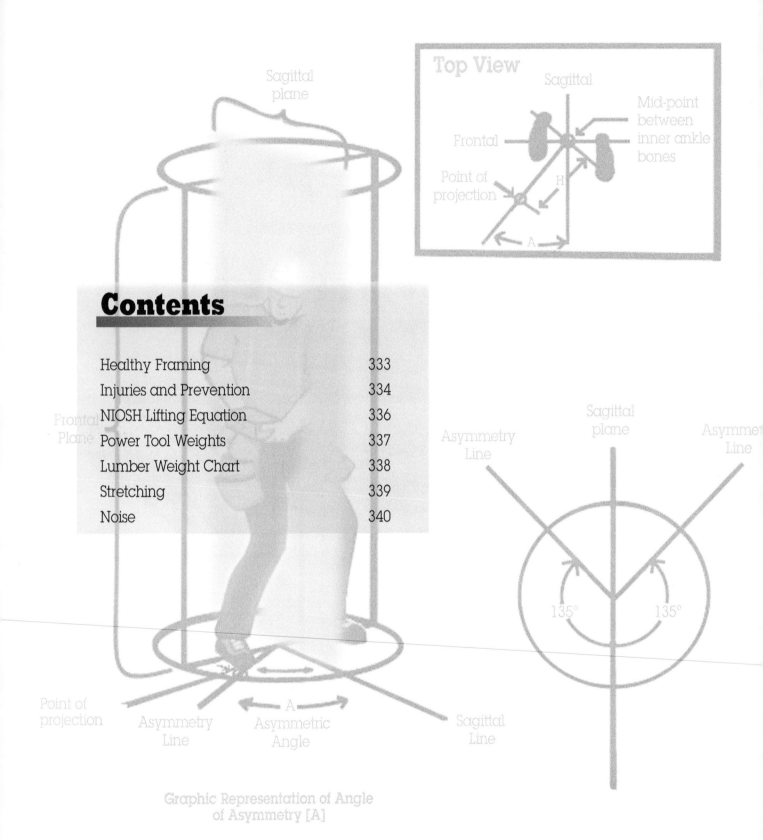

Contents

Graphic Representation of Angle of Asymmetry [A]

Chapter Sixteen

HEALTHY FRAMING

Healthy framing affects how you feel and your enjoyment of life. It is also good for preservation of income as a framer and for your future retirement.

In this chapter I will discuss various issues related to preservation of your health as a framer. Some of the topics such as diet may seem outside the scope of what you might expect in a framing book; however, I have found that diet along with the other topics that I discuss can affect your happiness and health. This chapter does not provide any solutions for preserving your health, but it does give you information, so you can make intelligent decisions on how you conduct your work life to maximize your preservation of health.

I have no medical training other than First Aid. I have consulted with doctors while writing; however, please do not consider this generalized information as medical advice.

This chapter includes the following:

1. A look at the typical injuries of framers.
2. Preventive suggestions for common injuries.
3. NIOSH Lifting Equation
4. Weight Chart for Lumber
5. Weight Chart for Tools
6. Stretching
7. Decibel Levels for Tools and Exposure
8. Diet

Growing up, we faced new tasks and challenges with an aggressive thought: do whatever I can, and then if need be, push my muscles to accomplish what I can't. The work environment, however, creates a different reality. The demands for strength and the needs for repetitive motion not only strain our physical well-being, but also expose us to possible harm. With framing in particular, it is good to maintain that attitude that you can do anything; however, because of the inherent possibility of overdoing certain physical activities, it is important to recognize the long-term effects on your body of maintaining this attitude.

Medical records of framers show where it has been common for them to exceed their body's natural ability. Back problems, tennis elbow, shoulder and kneecap bursitis, and carpal tunnel are some of the most common injuries.

The good news is that as someone with a physically demanding job, you will be achieving the benefits of a healthy active lifestyle without having to make daily trips to the gym.

To understand the factors that pose health risks, let's take a realistic look at what injuries framers sustain. The Bureau of Labor Statistics (BLS) of the US Department of Labor tracks occupational injuries and illnesses involving days away from work for framers. The information is not ideally categorized for framing injuries, but the information can give some direction for preventing injury.

According to the statistics, you are 10 times more likely to have an injury in the first 3 months as compared to the first 5 years, and 625 times more likely in the first 3 months as compared to the time after 5 years. The learning curve is apparent in these statistics, and what it tells us is that anything that we can learn before starting framing will probably help us in those first three risky months.

The BLS separates different types of exposures into categories.

One of the BLS categories, "Event or exposure," records that 35% of all injuries were being struck by an object, while 33% were from falls, slips, or trips, and 26% were from overexertion and bodily reaction.

A second category of the BLS is "Nature of injury, illness." Some of the major injuries or illnesses are Cuts, Lacerations, Punctures, 35%; Bruises, Contusions, 17%; Sprains, Strains, Tears, 14%; Factures, 9%; Soreness, Pain, 6%.

A third category is "Part of body affected." Some of the major body parts affected are Trunk, 26%; Hand, 21%; Head, 10%; Shoulder, 6%; Arm, 6%; Knee, 5%; Wrist, 3%; Ankle, 3%.

Another category that BLS tracks is "Source of injury, illness." Some of the major areas listed are Parts and Materials, 40%; Hand Tools, 16%; Ladders, 17%; Worker Motion or Position, 6%; Floors, Walkways, Ground Surfaces, 6%; Vehicles, 2%.

Injuries and Prevention

Unfortunately, the form OSHA 300 used for collecting this information is very general in nature; nevertheless, we can learn a lot from it. With this information in mind, the following looks at some specific injuries.

Musculoskeletal disorders are injuries and disorders that affect the human body's movement or musculoskeletal system, which includes muscles, tendons, ligaments, nerves, blood vessels, discs, etc. Some of the more common musculoskeletal disorders and injuries are back problems, tendinitis, rotator cuff tears, sprains, and strains. Following are parts of the body with common injuries and prevention suggestions.

Back

The BLS reports that the part of the body affected most by occupational injuries is the trunk at 26%. The two main concerns with the back are the back muscles and the discs, which are jellylike pads between the vertebrae. Sprains are injuries or tears to a ligament. Ligaments attach one vertebra to another. Strains are an injury to muscles that have been used or stretched too much. Muscles around strained and sprained ligaments are irritated by them, adding to the discomfort and pain.

Back Prevention Suggestions

1. Lift using the large muscles in your legs. Squat down and keep your back straight.

2. Review the HIOSH Lifting Equation (in this chapter) to get a good idea of how lifting on the job affects your body.

3. When material is delivered, have it delivered as near as possible to where you are going to use it.

4. Most lumber companies understand the order in which you use lumber and will stack the truckload of lumber so the lumber you use first is on top. It is usually worth the time to call them and go over their loading order.

5. When you must move material, place it in a position as close to where you are going to use it as possible and at a height that limits lifting. It also helps if you set it on something that allows you to get your fingers under for lifting.

6. Wear a low back brace to limit the bending of your back and aid in recovery.

7. Use an auto-feed screw gun with an extension if you are screwing a floor.

Hands

The second most affected part of the body is the hands. Tendinitis and De Quervain syndrome affect the hands. Caused by repeat movement of a joint, tendinitis is an inflammation and soreness in tendons.

Hand Prevention Suggestions

1. Help reduce the stress on your hands by keeping your tools sharp so that the pushing and pulling will be less.

2. Use handles for lifting materials (see "Handle for Lifting" picture)

3. Wear a wrist brace to limit motion and aid in recovery.

Elbow Brace
Back Brace
Wrist Brace
Knee Brace

Head

The head is the third most affected part of the body. Bruises, cuts, and punctures affect the head.

Head Prevention Suggestions

1. Good housekeeping. Keep your walkways clear of scrap and keep your cords and hoses untangled and aligned out of the way of being a tripping hazard.

2. Wear hard hats.

3. Don't leave tools on top of step ladders.

4. Meditation can assist with jobsite stress.

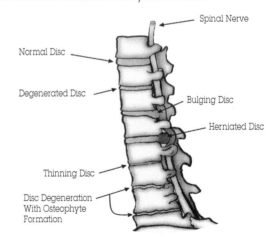

Spinal Nerve
Normal Disc
Degenerated Disc
Bulging Disc
Herniated Disc
Thinning Disc
Disc Degeneration With Osteophyte Formation

Shoulder

Shoulder injuries commonly occur from excess or overdoing of overhead work.

Shoulder Prevention Suggestions:

1. Use stilts for high work.
2. Use ladders or stools for high work.

Arm

A common injury of the arm is epicondylitis, or tennis elbow, which is a musculoskeletal disorder caused from strain on your elbow, chiefly from forceful twisting motions.

Arm Prevention Suggestions:

1. If you have a lot of screwing to do, use a power tool.
2. Wear an elbow brace for support and recovery.

Handle For Lifting

Knee

Knee muscles are connected to your legs by tendons. Small sacs of fluid between the tendons and bones are called bursae. The bursa lubricates the knee. If there is continual stress on the knee it can cause the bursa to get squeezed, stiff, inflamed, and swollen, which is known as bursitis. The stress can also cause the tendons in the knee to become inflamed and painful, known as tendinitis.

Knee Prevention Suggestions:

1. Wear knee pads
2. Wear a knee brace for support and recovery.

Wrist

Wrists present a common injury for framers in the form of carpal tunnel syndrome. A nerve and several tendons pass through the carpal tunnel in your wrist, which is surrounded by bones and tissue. Tendinitis, or swelling tendons, squeezes the nerve and often leads to pain, tingling, or numbness of your wrist, arm, or hand.

Wrist Prevention Suggestions:

1. Use your other hand where possible.
2. Try to limit any motion that causes pain.
3. Wear a wrist brace for support and recovery.

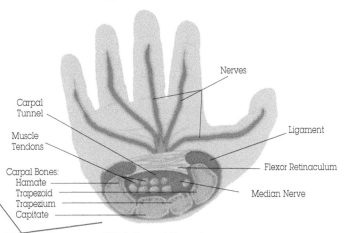

Wrist: Carpal Tunnel

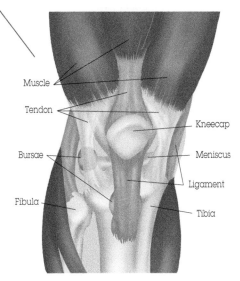

Knee: Muscles, Tendons, Ligaments, Bursa

335

NIOSH Lifting Equation

When you are out framing a wall, your first thought is how to finish the task. If your back or some other part of your system does not hurt, then you just keep on framing and don't give a thought to possible injuries. What you could be thinking about is how to integrate prevention into your work habits so that you don't have to think about it. For example, after you learn that you should lift using your leg muscles instead of your back muscles and you make it a habit to lift that way, then you will always lift with your leg muscles.

There is other lifting, however, that you can think about to protect your body. Probably the best information we have related to how different movements affect our body is the NIOSH Lifting Equation formula. NIOSH created this formula to distill the results of tests that have been conducted to calculate the severity of different movements on our bodies. The formula shows how much the weight of an object should be reduced to not increase the strain on the body from our increased movement. In other words, if you are lifting something, how much extra strain is put on your body as these different lifting factors are increased? The movements that are shown in this formula are the common movements that we should consider when we think about not overdoing lifting. They are not exact and cannot predict injury, but they can give you some information to think about as you attempt to balance your daily tasks with your desire to maintain good health.

The formula: **RWL** = LC × HM × VM × DM × AM × FM × CM
The factors are:
> **RWL** = Recommended Weight Limit
> **LC** = Load Constant
> **HM** = Horizontal Multiplier
> **VM** = Vertical Multiplier
> **DM** = Distance Multiplier
> **AM** = Asymmetric Multiplier
> **FM** = Frequency Multiplier
> **CM** = Coupling Multiplier

Don't let the multiplier divert you. Basically, all this formula says is that each of the factors can affect the strain on your body, and the formula tells you how much.

The following is a discussion of these seven multipliers:

1. **Load Constant** is set at 51 pounds, which is generally considered a maximum load nearly all healthy workers should be able to lift under best conditions for a continuous time period.

2. **Horizontal Multiplier** relates to the horizontal distance the object is from your body when you start lifting. If it is 10" or less from your body the weight does not need to be reduced. From 10" to 25" the recommended weight limit is reduced by a factor of 10 / H (see illustration) as part of the RWL formula. It is not recommended to lift anything that is further than 25" from your body.

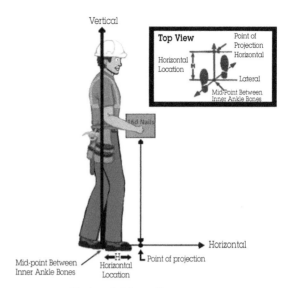

Horizontal Location

3. **Vertical Multiplier** relates to the effect of the vertical starting position of the lift, which increases as you go up or down from a height of 30 inches, or "knuckle height" for an average height worker of 66" (adjust for your own height). The formula is $(1 - (.0075 |V - 30|))$ and does not recommend starting the lift higher than 70 inches (see illustration).

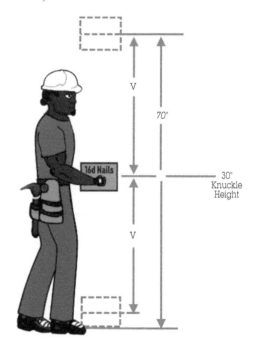

Vertical Location

4. **Distance Multiplier** relates to the lifting distance from a height of 30 inches, which is "knuckle height" for an average-height worker of 66" (adjust for your own height). The formula is $(.82 + (1.8 / D))$. It is recommended that from 0" to 10" no reduction in weight is needed, but weight should be decreased above 10" progressively until 70"; lifting above this height is not recommended (see illustration).

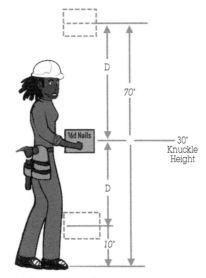

Distance Location

5. **Asymmetric Multiplier** is complicated, but know that the further you twist from lifting straight up and down with your body, the better it would be to reduce the weight. The formula is 1 – (.0032A). What it says is that the larger the angle from 1 degree away from straight, the less weight should be lifted. It is not recommended to lift further away than 135 degrees (see illustration).

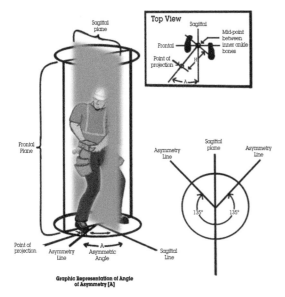

Graphic Representation of Angle of Asymmetry [A]

6. **Frequency Multiplier** is another factor that is complicated to fit into the overall RWL formula. It measures the effects of repeated movements of the body. It uses two factors: how many lifts occur per minute for a fifteen-minute period and how much time there is for recovery. It uses three time periods: short (less than an hour), medium (between one hour and two hours), and long (2 to 8 hours). In general, as frequency increases, the weight should decrease. Any more than 15 lifts per minute is not recommended. Recovery time is important in figuring this multiplier.

7. **Coupling Multiplier** measures the effect of being able to get a good grip on the object being lifted. There are three categories: Good, Fair, and Poor. The categories range

from a nice, square, balanced package with handles, to an irregular, bulky, hard-to-handle package with no handles. The worse the coupling, the less weight should be lifted.

An example of using handles are Vacuum Suction Cup Glass Lifters, which are attached to windows to provide handles (see below).

Vacuum Suction Cup Glass Lifters

This NIOSH Lifting Equation seems overly complicated; however, it is a very hard task to try and explain the effects of lifting on the body, and this is probably the best explanation available at the present. It can be used to further your understanding of the effects of lifting on your body and how you can adjust your lifting to preserve your body.

It takes many muscles working in many different ways to frame. It is difficult to understand coordination of all these muscles and movements to try and prevent injury. It is good, however, to understand the consistent variables that affect lifting. To supplement these, I have prepared charts of power tool weights and wood weights. These are not meant to be comprehensive, but rather to give you some general guidelines to use when thinking about tool purchase or planning your lifting tasks.

Power Tool Weights

TOOL	Average Weight, lb.
Circular Saw Worm Drive Corded	13
Circular Saw Worm Drive Cordless	12
Circular Saw Sidewinder Corded	9
Circular Saw Sidewinder Cordless	9
Recipro Saw Corded	10
Recipro Saw Cordless	9
Hammer Drill Corded	6
Hammer Drill Cordless	5
Drill/Driver Corded	6
Drill/Driver Cordless	4
Drill Corded	4
Drill Cordless	4
Stud/Joist Drill Corded	13
Stud/Joist Drill Cordless	14
Impact Wrench Corded	7
Impact Wrench Cordless	7
Router Corded	10
Air Compressor Wheelbarrow	167

Lumber Weight Chart

Lumber	Kiln Dried lbs/ft	Kiln Dried lbs/88-5/8"	Kiln Dried lbs/8ft	Kiln Dried lbs/12ft		Green lbs/ft	Green lbs/88-5/8"	Green lbs/8ft	Green lbs/12ft
			Stud					Stud	
2 x 4	1.29	9.53	10.32	15.48		1.66	12.26	13.28	19.92
2 x 6	2	14.77	16	24					
2 x 8	2.65	19.57	21.2	31.8					
2 x 10	3.33	24.59	26.64	39.96					
2 x 12	4.07	30.06	32.56	48.84					
4 x 4						3.6	26.59	28.8	43.2
4 x 6						5.38	39.73	43.04	64.56
4 x 8						7.19	53.1	57.52	86.28
4 x 10						9	66.47	72	108
6 x 6						8.1	59.82	64.8	97.2
6 x 8						11.04	81.53	88.32	132.48
6 x 10						13.99	103.32	111.92	167.88
6 x 12						15.5	114.47	124	186

Lumber	Pressure Treated lbs/ft	Pressure Treated lbs/88-5/8"	Pressure Treated lbs/8ft	Pressure Treated lbs/12ft		Thickness	Plywood Per Sheet		OSB Per Sheet
			Stud						
2 x 4	2.12	15.66	16.96	25.44					
2 x 6	3.19	23.56	25.52	38.28		3/8"	35.2		38.4
2 x 8	4.38	32.35	35.04	52.56		1/2"	48		54.4
2 x 10	5.5	40.62	44	66		5/8"	60.08		67.2
2 x 12	6.6	48.74	52.8	79.2		3/4"	73.6		80
4 x 4	4.79	35.37	38.32	57.48		1-1/8"	105.6		115.2
4 x 6	7	51.7	56	84					
4 x 8	9.32	68.83	74.56	111.84					
4 x 10	11.66	86.11	93.28	139.92					
4 x 12	14	103.39	112	168					

I-Joists	Depth	lbs/ft	lbs/10'	Flange
110 TJI	9-1/2"	2.3	23	1-3/4"
210 TJI	9-1/2"	2.6	26	2-1/16"
230 TJI	9-1/2"	2.7	27	2-5/16"
110 TJI	11-7/8"	2.5	25	1-3/4"
210 TJI	11-7/8"	2.8	28	2-1/16"
230 TJI	11-7/8"	3	30	2-5/16"
360 TJI	11-7/8"	3	30	2-5/16"
560 TJI	11-7/8"	4	40	3-1/2"
110 TJI	14"	2.8	28	1-3/4"
210 TJI	14"	3.1	31	2-1/16"
230 TJI	14"	3.3	33	2-5/16"
360 TJI	14"	3.3	33	2-5/16"
560 TJI	14"	4.2	42	3-1/2"
110 TJI	16"	3	30	1-3/4"
210 TJI	16"	3.3	33	2-1/16"
230 TJI	16"	3.5	35	2-5/16"
360 TJI	16"	3.5	35	2-5/16"
560 TJI	16"	4.5	45	3-1/2"

Lumber Weight Chart

Structural Lumber	Composite Grade	Lumber Width	Depth	lbs/ft	lbs/10ft
LSL	1.3E	3-1/2"	4-3/8"	4.5	45
LSL	1.3E	3-1/2"	5-1/2"	5.6	90
LSL	1.3E	3-1/2"	7-1/4"	7.4	74
LSL	1.55E	1-3/4"	9-1/2"	5.2	52
LSL	1.55E	1-3/4"	11-7/8"	6.5	65
LSL	1.55E	1-3/4"	14"	7.7	77
LSL	1.55E	1-3/4"	16"	8.8	88
LSL	1.55E	3-1/2"	9-1/2"	10.4	104
LSL	1.55E	3-1/2"	11-7/8"	13	130
LSL	1.55E	3-1/2"	14"	15.3	153
LSL	1.55E	3-1/2"	16"	17.5	175
LVL	2.0E	1-3/4"	5-1/2"	2.8	28
LVL	2.0E	1-3/4"	7-1/4"	3.7	37
LVL	2.0E	1-3/4"	9-1/4"	4.7	47
LVL	2.0E	1-3/4"	9-1/2"	4.8	48
LVL	2.0E	1-3/4"	11-1/4"	5.7	57
LVL	2.0E	1-3/4"	11-7/8"	6.1	61
LVL	2.0E	1-3/4"	14"	7.1	71
LVL	2.0E	1-3/4"	16"	8.2	82
LVL	2.0E	1-3/4"	18"	9.2	92
LVL	2.0E	1-3/4"	20"	10.2	102
PSL	2.0E	3-1/2"	9-1/4"	10.1	101
PSL	2.0E	3-1/2"	9-1/2"	10.4	104
PSL	2.0E	3-1/2"	11-1/4"	12.3	123
PSL	2.0E	3-1/2"	11-7/8"	13	26
PSL	2.0E	3-1/2"	14"	15.3	153
PSL	2.0E	3-1/2"	16"	17.5	175
PSL	2.0E	3-1/2"	18"	19.7	197
PSL	2.0E	5-1/4"	9-1/4"	15.2	152
PSL	2.0E	5-1/4"	9-1/2"	15.6	156
PSL	2.0E	5-1/4"	11-1/4"	18.5	185
PSL	2.0E	5-1/4"	11-7/8"	19.5	195
PSL	2.0E	5-1/4"	14"	23	230
PSL	2.0E	5-1/4"	16"	26.3	263
PSL	2.0E	5-1/4"	18"	29.5	295
PSL	2.0E	7"	9-1/4"	20.2	202
PSL	2.0E	7"	9-1/2"	20.8	208
PSL	2.0E	7"	11-1/4"	24.6	246
PSL	2.0E	7"	11-7/8"	26	260
PSL	2.0E	7"	14"	30.6	306
PSL	2.0E	7"	16"	35	350
PSL	2.0E	7"	18"	39.4	394

Stretching

Stretching improves muscle flexibility. Many larger construction companies have implemented daily stretching programs at the beginning of each day. These programs express a positive attitude by the company toward the framer's health, but studies have found varying results on their ability to prevent musculoskeletal disorders and injuries. One of the most respected studies by the Department of Occupational and Environmental Safety and Health of the University of Wisconsin, states, "While research does support that stretching improves flexibility/ROM and self-worth, stretching alone might not prevent work-related musculoskeletal disorders and injuries."

Stretching Basics

1. Hold each stretch for 15 to 30 seconds.
2. Stretch slowly, and do not bounce.
3. Stretch both right and left sides.
4. Breath while stretching.
5. Stretch regularly.
6. Stretch your hands, arms, shoulders, back, neck, legs, chest, and wrists.
7. Stop the stretch if you experience pain.

Noise

Hearing loss is a concern for framers. I did not pay attention when I was framing, and now, retired, I have tinnitus (ringing in the ears). OSHA (Occupational Safety and Health Administration) sets limits on safe levels of noise. Below is a chart showing the safe decibel levels, along with the decibel levels for common framing tools. If you are in a condition where the sound levels are higher than the permissible noise exposure, then you need to wear ear protection (see Noise Protection below).

Permissible Noise Exposure

Duration Per Day Hours (T)	Sound Level dBA slow response
8	90
6	92
4	95
3	97
2	100
1-½	102
1	105
½	110
¼ or less	115

Power Tool Decibel Levels

Tool	Decibels
Circular Saw	108
Reciprocating Saw	106
Drill	92
Mitre Saw	107
Hammer Drill	104
Jig Saw	98
Impact Wrench	107
These are average decibels levels from numerous tools while loaded (in use).	

OSHA has a formula to determine if you need to wear ear protection to reduce noise. The formula states that if you add up all the noise exposure that you had in an eight-hour day it should not be greater than 1. To calculate the formula, just figure the noise exposure you had during the day and divide the permissible duration limit for the day from the chart above by the actual exposure time. For example, if you operated a drill for more than 6 hours, you should wear ear protection the whole time. Drill = 92 Decibels = 6 hours (permissible hours per day). T = % = 1. Another example: A circular saw at 108 decibels would allow approximately 48 minutes (between 1 hour at 105 decibels and ½ hour at 110 decibels). So, if you use a circular saw for 24 minutes and a drill more than 3 hours in the same day you will need ear protection.

$F = T_1/L_1 + T_2/L_2 \ldots \ldots <1$
F = Noise exposure factor
T = Duration of time allowed (see "Permissible Noise Exposure")
L = Duration or length of time using tool

It is a little complicated, but if you just get an idea of how much noise you can listen to during a day before you need to wear protection, it can direct you in your decision-making process.

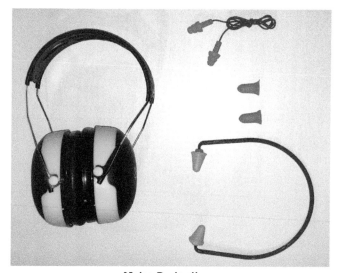

Noise Protection

Diet

Diet is one of the subjects that I hesitate to even mention, lest I become one of the numerous snake oil doctors out there. What we put in our bodies has a major impact on our health and our productivity as a framer. I do not want to suggest a diet, but it is good to eat a healthy balanced diet that provides you with the nourishment that will allow your muscles to perform to their maximum capacity.

GLOSSARY, SPANISH FRAMING TERMS, AND INDEX

GLOSSARY

anchor bolt
A bolt for connecting wood members to concrete or masonry.

backer
Three studs nailed together in a U-shape to which a partition is attached.

bargeboard
A board attached to a gable rafter to which a rake board is attached.

beam
A large, horizontal structural member of wood or steel, such as a girder, rafter, or purlin.

bearing wall
A wall that supports the load of the structure above it.

bevel
Any angle that is not 90°. Also, a tool for marking such an angle.

bird's mouth
The notch in a rafter that rests on the top plate of a wall.

blocking
Small pieces of wood used to secure, join, or reinforce members, or to fill spaces between members.

box nail
Nail similar to a common nail, but with a smaller diameter shank.

bridging
Bracing installed in an X-shape between floor joists to stiffen the floor and distribute live loads. Also called cross-bridging.

building line
The bottom measurement of a rafter's run (the top being the ridge line); the plumb cut of the bird's mouth.

camber
A slight crown or arch in a horizontal structural member, such as beam or truss, to compensate for deflection under a load.

cantilever
Any part of a structure that projects beyond its main support and is balanced on it.

cant strip
A length of lumber with a triangular cross-section used around edges of roofs or decks to help waterproof.

casing
A piece of wood or metal trim that finishes off the frame of a door or window.

casing nail
A large finish nail with a small, cone-shaped head.

chalk line
A string covered with chalk used for marking straight lines.

check
A small split in a piece of lumber that runs parallel to the grain.

cheek cut (side cut)
An angle cut that is made to bear against another rafter, hip, or valley.

chord
Any principal member of a truss system. In a roof truss, the top chord replaces a rafter, and the bottom chord replaces a ceiling joist.

circumference
The perimeter measurement of a circle.

collar beam
A horizontal board that connects pairs of rafters on opposite roof slopes.

column
A vertical structural member.

common nail
Nail used in framing and rough carpentry, having a flat head about twice the diameter of its shank.

common rafter
A rafter running from a wall directly to a ridge board.

connection angle
The angle at the end of a rafter needed to connect to other rafters, hips, valleys, or ridge boards.

connectors
Beams, or construction hardware specifically designed for common framing connections.

corner
Two studs nailed together in an L-shape, used to attach two walls and provide drywall backing.

cornice
The horizontal projection of a roof overhang at the eaves, consisting of lookout, soffit, and fascia.

crawl space
The area bounded by foundation walls, first-floor joists, and the ground in a house with no basement.

cricket
A small, sloping structure built on a roof to divert water, usually away from a chimney. Also called a saddle.

cripple
A short stud installed above or below a horizontal member in a wall opening.

cripple jack rafter
A rafter that runs between a hip rafter and a valley rafter.

crown
The high point of a piece of lumber that has a curve in it.

cup
Warp across the grain.

d
Abbreviation for penny (nail size). The abbreviation comes from the Roman word denarius, meaning coin, which the English adapted to penny. It originally referred to the cost of a specific nail per 100. Today it refers only to nail size.

dead load
The weight of a building's construction materials, in place, in pounds per square foot. This includes walls, floors, roofs, ceilings, stairways, finishes, HVAC and plumbing systems, and other architectural and structural items and building systems.

diagonal
The distance between the far point on the run and the high point on the rise. (Similar to hypotenuse in mathematical terms.)

diagonal percent
The diagonal divided by the run, used when cutting rafters.

dormer
A structure with its own roof projecting from a sloping roof, typically to accommodate a vertical window.

double cheek cut
A two-sided cut that forms a V at the end of some rafters, especially in hip and gambrel roofs.

dry line
A string line (as opposed to a chalk line) used to establish a straight line.

eave
The part of a roof that projects beyond its supporting walls.

engineered wood products (EWP)
Building framing, joists, and beams made from wood strands or fibers held together with a binder.

face-mount hangers
Hangers that nail onto the face or vertical surfaces of their supporting members.

face nailing
Nailing at right angles to the surface.

fascia
A vertical board nailed to the lower ends of rafters that form part of a cornice.

fireblock
A short piece of framing lumber nailed horizontally between joists or studs to partially block the flow of air and, thus, to slow the spread of fire.

footings
The base, usually poured concrete, on which the foundation wall is built. The footing and foundation wall are often formed and poured as a single unit. It spreads and transmits the load directly to the soil.

footprint
The area that falls directly beneath and shares the same perimeter as a structure.

form
A mold of metal or wood used to shape concrete until it has set.

foundation
The building's structural support below the first-floor construction. It rests on the footing, and transfers the weight of the building to the soil.

frame
The skeleton of a building. Also called framing.

framer
A person who performs rough carpentry, building the frame of the structure to which sheathing and finish treatments will be attached.

framing anchor
A metal device for connecting wood framing members that meet at right angles.

framing point
The point where the center lines of connecting rafters, ridges, hips, or valleys meet.

furring
Strips of wood fastened across studs or joists to a level or plumb nailing surface for finish wall or ceiling material, usually drywall. Also called strapping.

gable
The triangular part of an end wall between the eaves and ridge of a house with a peaked roof.

gable roof
A roof shape characterized by two sections of roof of constant slope that meet at a ridge; peaked roof.

gambrel roof
A roof shape similar to a gable roof, but with two sections of roof on each side of the ridge, the lower section being steeper than the upper.

girder
A primary horizontal beam of steel or wood used to support other structural members at isolated points along its length.

glu-lam beams
Structural beams created by gluing 2× dimensional lumber together in a structurally ordered pattern.

grade
1. A designation of quality, especially of lumber and plywood.

2. Ground level. Also the slope of the ground on a building site.

gusset
A flat piece of plywood or metal attached to each side of two framing members to tie them together, or strengthen a joint.

gusset plates
Plates fastened across a joint that can be used to replace a ridge board.

header
Any structural wood member used across the ends of an opening to support the cut ends of shortened framing members in a floor, wall, or roof.

hip
The outside angle where two adjacent sections of roof meet at a diagonal. The opposite of a valley.

hip diagonal (or valley diagonal)
The distance between the far point on the hip or valley run and the high point on the hip or valley rise.

hip rafter
The diagonal rafter, which forms a hip.

hip roof
A roof shape characterized by four or more sections of constant slope, all of which run from a uniform eave height to the ridge.

hip run (or valley run)
The horizontal distance below the hip or valley of a roof, from the outside corner of the wall to the center framing point.

hip-val diagonal percent
The hip or valley diagonal divided by the hip or valley run.

hold-downs
Connections used to transfer tension loads between floors. Commonly used for foundations, wall-to-wall connections, wall-to-concrete connections, and wall or floor-to-drag strut. Also called anchor downs or tie-downs.

I-joist
An engineered wood product created with two flanges joined by a web and that develops certain structural capabilities. I-joists are also used for rafters.

***International Building Code* (IBC)**
The building code established in 2000 by the International Code Council (ICC) to cover all buildings other than one- and two-family dwellings and multiple single-family dwellings not more than three stories in height.

***International Residential Code* (IRC)**
The International Code Council's (ICC) building code that covers all one- and two-family dwellings and multiple single-family dwellings not more than three stories in height.

jack rafter
A short rafter, usually running between a top plate and a hip rafter, or between a ridge and a valley rafter.

jack stud
A shortened stud supporting the header above a door or window. Also called a trimmer or jamb stud.

jamb
The side of a window or door opening.

joint
The line along which two pieces of material meet.

joist
One of a parallel series of structural members used for supporting a floor or ceiling. Joists are supported by walls, beams, or girders.

joist hanger
A metal framing anchor for holding joists in position against a rim joist, header, or beam.

kerf
The cut made by a saw blade.

king stud
A vertical support member that extends from the bottom to top plate alongside an opening for a door or window.

kneewall
A short wall under a slope, usually in attic space.

layout language
The written words and symbols the lead framer uses to communicate directions for framing work.

lead framer
Foreman or leader of a framing crew.

ledger
A strip of lumber attached to the side of a girder near its bottom edge to support joists. Also, any similar supporting strip.

level
Perfectly horizontal. Also refers to various types of measuring instruments used to determine horizontal or vertical alignments.

lightweight plate
A plate that goes under the bottom plate to raise walls for lightweight concrete or gypcrete.

live load
The total variable weight in pounds per square foot on a structure including occupants, furnishings, snow, and wind.

load path
The path taken by artificial and natural forces that affect a building when they create a load exerted on the building.

lookout
A horizontal framing member attached to a rafter, supporting an overhanging portion of the roof.

lumber
Wood cut at a sawmill into usable form.

makeup
Parts of a wall (such as backers, corners, headers, and stud trimmers) cut before the wall is spread.

mansard roof
A type of roof with two slopes on each of four sides, the lower slope much steeper than the upper and ending at a constant eave height.

mortise
A recess cut into wood.

mudsill
The lowest plate in a frame wall that rests on the foundation or slab.

nail gun
A hand-operated tool powered by compressed air that drives nails.

nominal size
The rounded-off, simplified dimensional name given to lumber. For example, a piece of lumber whose actual size is 1-½" × 3-½" is given the more convenient, nominal designation of 2" × 4".

nonbearing wall
A dividing wall that is not designed to support the load of any structure above it.

nosing
The rounded front edge of a stair tread that extends over the riser.

on center (O.C.)
Layout spacing designation that refers to the distance from the center of one framing member to the center of another.

oriented strand board (OSB)
Wood made out of flakes, strands, or wafers sliced from small wood logs bonded under heat with a waterproof and boil-proof resin binder.

overhang
The part of a roof that extends beyond supporting walls.

overhang diagonal
The distance between the far point on the overhang run and the high point on the overhang rise.

overhang hip run
The horizontal distance below the hip or valley of a roof, from the outside corner of the wall to the center framing point.

parallel
Extending in the same direction and equidistant at all points.

parapet
A low wall or rail at the edge of a balcony or roof.

particleboard
Engineered lumber made of wood particles bonded by an adhesive under a heat and pressure, and formed into a solid, three-layered panel with two surface layers. (Usually used as an underlayment.)

partition
An interior wall that divides a building into rooms or areas.

party wall
A wall between two adjoining living quarters in a multi-family dwelling.

penny
Word applied to nails to indicate size; abbreviated as "d."

perpendicular
At right angles to a plane or flat surface.

pitch
The slope of a roof.

pitch angle
The vertical angle on the end of a rafter that represents the pitch of a roof.

plate
A horizontal framing member laid flat.

    bottom plate
    The lowest plate in a wall in the platform framing system, resting on the subfloor, to which the lower ends of studs are nailed.

    double plate
    The uppermost plate in a frame wall that has two plates at the top. Also called a cap or rafter plate.

    sill plate
    The structural member, attached to the top of the foundation, that supports the floor structure. Also called a sill or mudsill.

top plate
The framing member nailed across the upper ends of studs and beneath the double plate.

platform framing
A method of construction in which wall framing is built on and attached to a finished box sill. Joists and studs are not fastened together as in balloon and braced framing.

plumb
Straight up and down, perfectly vertical.

plumb and line
The process of straightening all the walls so they are vertical and straight from end to end.

plumb cut
Any cut in a piece of lumber, such as at the upper end of a common rafter, that will be plumb when the piece is in its final position.

purlin
The horizontal framing members in a gambrel roof between upper and lower rafters.

Pythagorean theorem
The theorem that the sum of the squares of the lengths of the sides of a right triangle is equal to the square of the length of the hypotenuse.

rabbet
A groove cut in or near the edge of a piece of lumber to receive the edge of another piece.

rafter
One of a series of main structural members that form a roof.

rake
The finish wood member running parallel to the roof slope at the gable end.

rake wall
Also called gable end walls. Any wall that is built with a slope.

reveal
The surface left exposed when one board is fastened over another; the edge of the upper set slightly back from the edge of the lower.

ridge
The horizontal board to which the top ends of rafters are attached.

ridge end rafter
A rafter that runs from the end of a ridge.

rim joist
A joist that forms the perimeter of a floor framing system.

rise
1. In a roof, the vertical distance between the top of the double plate and the point where a line, drawn through the edge of the double plate and parallel to the roof's slope, intersects the center line of the ridgeboard.

2. In a stairway, the vertical height of the entire stairway measured from floor to floor.

riser
The vertical board between two stair treads.

roof sheathing
Material, usually plywood, laid flat on roof trusses or rafters to form the roof.

rough opening (R.O.)
Any opening formed by the framing members to accommodate doors or windows.

rout
To cut out by gouging.

run
1. In a roof with a ridge, the horizontal distance between the edge of the rafter plate (building line) and the center line of the ridge board.

2. In a stairway, the horizontal distance between the top and bottom risers plus the width of one tread.

scaffold
The covering (usually wood boards, plywood, or wallboards) placed over exterior studding or rafters of a building; provides a base for the application of exterior wall or roof cladding. Any temporary working platform and the structure to support it.

scribe
To mark for an irregular cut.

seat cut
The horizontal cut in a bird's-mouth that rests on the double plate.

shear wall
A wall that, in its own plane, carries shear resulting from forces such as wind, blast, or earthquake.

sheathing
Material, usually plywood, attached to studs to form the outside wall and provide structural strength.

shed roof
A roof that slopes in only one direction.

shim
A thin piece of material, often tapered (such as a wood shingle) inserted between building materials for the purpose of straightening or making their surfaces flush at a joint.

sill
1. A sill plate.

2. The structural member forming the bottom of a rough opening for a door or window. Also, the bottom member of a door or window frame.

single cheek cut
A bevel cut at the end of a rafter, especially in hip and gambrel roofs.

sleeper
Lumber placed on a concrete floor as a nailing base for wood flooring.

slope
The pitch of a roof, expressed as inches of rise per twelve inches of run.

soffit
The underside of a projection, such as a cornice.

solid bridging
Blocking between joists cut from the same lumber as the joists themselves and used to stiffen the floor.

spacer
Any piece of material used to maintain a permanent space between two members.

span
The distance between structural supports, measured horizontally (typically from the outside of two bearing walls).

speed square
A triangle-shaped tool used for marking perpendicular and angled lines.

square
1. At 90°, or a right angle.

2. The process of marking and cutting at a right angle.

3. Any of several tools for marking at right angles and for laying out structural members for cutting or positioning.

4. A measure of roofing and some siding materials equal to 100 square feet of coverage.

squash blocks
A structural block used to support point loads.

stair
A single step or a series of steps or flights of steps connected by landings, used for passage from one level to the next.

stair nuts
Two screw clamps that are attached to a framing square for marking stair stringers.

stair stringer
An inclined board that supports the end of the treads. Also known as stair jacks.

stairway
A flight of stairs, made up of stringers, risers, and treads.

stairwell
The opening in a floor for a stairway.

stickers
Strips of scrap wood used to create an air space between layers of lumber.

stick frame
Method of framing involving building one structural member at a time, e.g. nailing one stud at a time in place.

stop
In general, any device or member that prevents movement.

story pole
A length of wood marked off and used for repetitive layout or to accurately transfer measurements.

stringer
In stairway construction, the diagonal member that supports treads and to which risers are attached.

structural
Adjective generally synonymous with "framing."

stud
The main vertical framing member in a wall to which finish material or other covering is attached.

subfloor sheathing
The rough floor, usually plywood, laid across floor joists and under finish flooring.

tail
The part of a rafter that extends beyond the double plate.

tail joist
A shortened joist that butts against a header.

tape
A measure of coiled, flexible steel.

template
A full-sized pattern.

threshold
The framing member at the bottom of a door between the jambs.

toenailing
Driving a nail at an angle to join two pieces of wood.

tongue
The shorter and narrower of the two legs of a framing square.

top-flange hangers
Hangers with flanges that attach to the top of supporting members.

transit line
In surveying, any line or a survey traverse that is projected, either with or without measurement, by the use of a transit or similar device.

tread
The horizontal platform of a stair.

trimmer
The structural member on the side of a framed rough opening used to narrow or stiffen the opening. Also, the shortened stud (jack stud) that supports a header in a door or window opening.

truss
An assembly for bridging a broad span, most commonly used in roof construction.

trussed joist
A joist in the form of a truss.

utility knife
A hand-held knife with a razor-like blade, commonly used to cut drywall, sharpen pencils, etc.

valley
The inside angle where two adjacent sections of a roof meet at a diagonal. The opposite of a hip.

valley rafter
A rafter at an inside corner of a roof that runs between and joins with jack rafters that bear on corner walls.

wall puller
A tool used for aligning walls.

wall sheathing
Material, usually plywood, attached to studs to form the outside wall and provide structural strength.

wane
A defect in lumber caused at the mill by sawing too close to the outside edge of a log and leaving an edge either incomplete or covered with bark.

warp
Any variation from straight in a piece of lumber; bow, cup, crook, or twist.

web stiffeners
A piece of wood or composite wood used to provide additional strength for the webs of I-joists.

western framing/platform framing
A framing system in which the vertical members are only a single story high, with each finished floor acting as a platform upon which the succeeding floor is constructed.

worm-drive saw
A circular power saw turned by a worm-gear drive. It is somewhat heavier and produces more torque on the blade than a standard circular saw.

SPANISH FRAMING TERMS

add (v)
agregar

addition
ampliación
expansión

adhesive
pegamento

air compressor
compresor de aire

align (v)
alinear

anchor
anclaje
anclar

anchor blocks
macizos
bloques de anclaje

anchor bolts
pernos de anclaje
tornillos de anclaje

angle
ángulo

angle (v)
angular

architect
arquitecto

area
área

assemble (v)
armar/ensamblar

attic
tapanco
entrepiso

backing
soporte
respaldo

balance (v)
balancear

balcony
balcón
terraza

basement
sótano

beam
viga

beam hanger
colgador/suspensor de viga

bend (v)
doblar

bevel
bisel

blind nailed
con clavos ocultos

block, blocking
trabas
trabar
bloque
bloquear

board
panel
tabla
tablero

bolt
perno
tornillo

box frame
bastidor de cajón

box nail
clavo para madera
clavo de cabeza grande plana

brace
tirante

brace and bit
taladro de mano
berbiquí y barrena

braced frame
estructura arriostrada
pórtico arriostrado

bracing
arriostramiento

bracket
brazo
soporte
ménsula

break (v)
romper

bridging
puntales de refuerzo

brush (v)
cepillar

bucket
cubeta
balde

foundation
fundación
cimentación

foundation bolt
anclaje de cimentación

foundation sill plate
placa de solera de fundación

frame
marco
estructura
pórtico
bastidor
armazón

frame, door/window
marco de puerta/ventana

framed
armado

framework
armazón

framing
estructura
intramado

framing square
escuadra

front
frente

furred out, furring
enrasado

gable
hastial

gable construction
construcción a dos aguas

gable rake
cornisa inclinada

gable roof
techo a dos aguas

garage
garaje
cochera

general contractor
contratista general

girder
viga maestra
viga principal
viga
jácena

girder supports/girder posts
apoyos de viga

grade
nivel de terreno
grado

grade beam
viga de fundación

graded lumber
madera elaborada
madera clasificada

grain
veta

groove
ranura

gross area
área total
área bruta

gypsum board
panel de yeso
tablero de yeso
plancha de yeso
plafón de yeso

hacksaw
sierra para cortar metales

hallway
pasillo

hammer
martillo

hammer (v)
martillar

handle
manija
mango
brazo
agarradera

handle (v)
manipular

hand saw
serrucho de mano

hang (v)
colgar

hangers
ganchos
colgaderos

hanging scaffold
andamio colgante

hard hat
casco de seguridad

hardware
herrajes
ferretería

hardwood
madera dura

header
cabezal
dintel
colector

head joint
junta vertical

height
altura

hinge
bisagra

hip
lima
lima hoya
lima tesa

hip roof
techo a cuatro aguas

hold (v)
sujetar/sostener

hold-down anchor
ancla de retención

hole
hoyo
agujero
boquete

hook (v)
enganchar

horizontal bracing system
sistema de arriostramiento
horizontal

hose
manguera

house
casa

impact
impacto

inch
pulgada

incline
declive
inclinación
ladera
pendiente

incline (v)
inclinar
ladear

install (v)
instalar

interior room
cuarto interior

interlocking
enclavamiento

International Building Code
Código Internacional para
Instalaciones

jamb
jamba
quicial

job site
lugar de la obra/en la obra
sitio de construcción
sitio del trabajo

join (v)
unir

joint
unión
junta
pegasón

joist
vigueta
viga

joist anchor
anclaje de vigueta

joist, end
vigueta esquinera

joist, floor
vigueta del piso

joist hanger
estribo para vigueta

junction
empalme
unión

king post
poste principal
columna

kitchen
cocina

knife, utility
cuchillo utilitario

ladder
escalera
escalera de mano

laminated wood
madera laminada

landing (stair)
descanso de escaleras

layout
croquis
diseño

lean (v)
apoyar/inclinar

ledge
retallo

level
nivel

lift (v)
leventar
elevar

line and grade
trazar y nivelar

load-bearing joist
viga de carga

lower (v)
bajar

lumber
madera de construcción
madera elaborada

measure (v)
medir

mitre box
caja de corte a ángulos

mitre cut
corte de inglete

mitre saw
sierra de retroceso para ingletes

nail (v)
clavar

nail gun
clavadora automática

nailing, face
con clavos sumidos

nailing strip
listón para clavar

nail puller
sacaclavos

nails
clavos

nail set
botador/embutidor de clavos

nonbearing wall
pared sin carga

nonflammable
no inflamable

nosings
vuelos

nut
tuerca

on center
de centro a centro

open (v)
abrir

opened
abierto

opening
abertura

overhang
voladizo
velo
alero

overlap
traslape
sobresolape
superposición
traslapo

owner
propietario

parallel
paralelo

particleboard
madera aglomerada

partition
tabique
separación
vivisión
partición

partition wall
pared divisoria

perimeter
perímetro

perpendicular
perpendicular

Phillips (head screwdriver)
desarmador de punta de cruz

pick-up truck
camioneta

pitch
pendiente

plan
plano

plane
cepillo

plank
tablón

plumb (v)
plomar

plumb line
hilo de plomada

plywood
madera prensada
tableros de madera prensada
chapeado
triply
madera terciada

pounds per square inch (PSI)
libras por pulgadas cuadradas

power supply
fuente de alimentación

prefabricate (v)
prefabricar

prefabricated
prefabricado

prehung door
puerta premontada

pressure-treated wood
tratada a presión

project
proyecto

property
propiedad
parcela

pull (handle, grip)
agarradera

pull (v)
jalar

push (v)
empujar

quality
calidad

quantity
cantidad

quote (price)
cotización

radial arm saw
serrucho guillotina

radial saw
sierra fija

rafter
cabrio
cabio

rail
cremallera
baranda
barandilla

raise (v)
subir

rake
rastrillo

ramp
rampa

rate of rise
velocidad de incremento

rating
clasificación

reach (v)
alcanzar

recessed
empotrado

reciprocating saw
sierra alternativa

reinforcement
refuerzo
armadura

ridge
cresta
cumbrera

ridge board
tabla de cumbrera

rim
borde

riser
contrahuella

rolling scaffold
andamio movible

roof
techo

roof, flat
azotea

roof, sloped
techo en pendiente

roof framing
armazón de tejado

roofing
techado

roofing square
cadro de cubierta de techo

roof sheathing
entarimado de tejado

roof truss
armadura de cubierta

room
cuarto
sala
habitación

rot
podredumbre

rough-in
instalación en obra negra/gruesa

safety
seguridad industrial

safety glasses
gafas de seguridad

sand (v)
lijar

sander
lijadora

sandpaper
lija

sash
marco de ventana

saw
sierra
serrucho

saw (v)
serruchar

saw, electric
sierra eléctrica

saw, hack
sierra para metales

saw, hand
serrucho de mano

saw, power
sierra eléctrica

sawhorse
burro

sawn timber
maderos aserrados

scaffold
andamio

scale (v)
escalar
desescamar

schedule
horario

scope
alcance

score (v)
marcar

screw
tornillo

screw (v)
atornillar

screw connector
conector con tornillo

screwdriver
destornillador
desarmador

screwed fitting
acoplamiento roscado

shear wall
muro cortante
muro de corte
muro sismorresistente

sheathing
entablado

shim
calza

side-hinged door
puerta con bisagras laterales

sill
soporte

sill plate
solera inferior

site
sitio

skylight
tragaluz
claraboya
lucernario

slope
pendiente
talud
declive
vertiente

soffit
sofito

sole plate
placa de base

span
luz
vano
claro

square
escuadra

squeeze (v)
apretar

stair, landing
rellano

stair, riser
contrahuella

stair, tread
huella

stairs
escaleras

stairway
escalera

stop (v)
parar

story
piso

strapping
flejes

stress
esfuerzo

structure
estructura

stud
montante
parante
barrote

stud anchors
barras de anclaje

stud bearing wall
muro portante con montante

stud walls
muros con montantes
paredes de barrotes

superintendent
superintendente

supervisor
supervisor

support
apoyo
soporte

support (v)
resistir
sostener
soportar
apoyar

table saw
sierra fija
sierra de mesa
sierra circular de mesa

tape measure
cinta de medir

termite
termita

tie
amarra
ligadura
tirante

tie (v)
amarrar

timber
maderos
madera de construcción

toenail
clavo oblicuo

tongue and groove
machihembrado

tool belt
cinturón de herramientas

tool box
caja de herramientas

tools
herramientas

tread
huella
peldaño

truss
cercha
reticulado
armadura
cabreada
caballete

valley (roofing)
limahoya

valley jack
cabio de lima hoya

wall
muro
pared
barda

window
ventana

wood
madera

yard
yarda

INDEX